Ins

Springer
*Tokyo
Berlin
Heidelberg
New York
Barcelona
Budapest
Hong Kong
London
Milan
Paris
Santa Clara
Singapore*

F. Dunn (Editor-in-Chief)
M. Tanaka, S. Ohtsuki, Y. Saijo (Editors)

Ultrasonic Tissue Characterization

With 161 Figures, Including 26 in Color

 Springer

Editor-in-Chief:
Floyd Dunn, Ph.D.
Professor of Electrical Engineering, Biophysics, and Bioengineering,
University of Illinois, 1406 West Green Street, Urbana, IL 61801, USA

Editors:
Motonao Tanaka, M.D.
Professor of Medicine, Institute of Development, Aging and Cancer, Tohoku University,
4-1 Seiryo-machi, Aoba-ku, Sendai, 980 Japan

Shigeo Ohtsuki, Ph.D.
Professor of Electrical Engineering, Precision and Intelligence Laboratory,
Tokyo Institute of Technology, 4259 Nagatsuta, Midori-ku, Yokohama, 226 Japan

Yoshifumi Saijo, M.D.
Institute of Development, Aging and Cancer, Tohoku University, 4-1 Seiryo-machi,
Aoba-ku, Sendai, 980 Japan

ISBN 4-431-70162-1 Springer-Verlag Tokyo Berlin Heidelberg New York

Printed on acid-free paper

© Springer-Verlag Tokyo 1996
Printed in Hong Kong
This work is subject to copyright. All rights are reserved, whether the whole or part of the material is concerned, specifically the rights of translation, reprinting, reuse of illustrations, recitation, broadcasting, reproduction on microfilms or in other ways, and storage in data banks.
The use of registered names, trademarks, etc. in this publication does not imply, even in the absence of a specific statement, that such names are exempt from the relevant protective laws and regulations and therefore free for general use.
Product liability: The publisher can give no guarantee for information about drug dosage and application thereof contained in this book. In every individual case the respective user must check its accuracy by consulting other pharmaceutical literature.

Printing & binding: Permanent Typesetting & Printing, Co., Ltd., Hong Kong

to Yoshimitsu Kikuchi
— who showed us the way to proceed

Preface

There are two major reasons for having this symposium. First, Tohoku University is the place where ultrasound investigations in Japan originated. Starting from the research studies of Professors H. Nukiyama and Y. Kikuchi of Tohoku University, Professor J. Saneyoshi of Tokyo Institute of Technology, and Dr. K. Kato of Osaka University – all graduates of Tohoku University – the results spread to all parts of Japan. More recently we have had acoustic macroscopic studies by researchers like Professor N. Chubachi. As regards tissue characterization, which was the main theme of the symposium, the collaboration among research workers in Japan and the United States started 10 years ago between Professor F. Dunn of the University of Illinois and staff members of Tohoku University and the Tokyo Institute of Technology. So this conference commemorates the 10th anniversary of that joint research effort.

The second reason for this conference is that the application of ultrasound has become wide spread and indispensable in the routine clinical activities of medicine. But there have not been many breakthroughs in terms of quantitative and qualitative measurement of the living body tissues. Also, there are many problems with regard to practical application. There are various points that have not been elucidated yet as to the physical and acoustical characteristics of ultrasound itself. The methodology has not in all cases been well established. Therefore, the scientific elucidation of these areas is essential. But we have very few investigators in the world specializing in this field. For further development of these kinds of investigations, global joint research might be necessary, and we think it is very important to have this kind of opportunity at least once every two or three years to bring practicing experts together for free discussion of current problems and future directions. We hope that this conference will serve to provide these opportunities.

M. Tanaka, M.D.

Introductory Remarks

Tissue characterization is a term that has been used in medical diagnostics for a very, very long time. Its introduction into the field of medical ultrasound was more than 20 years ago, prior to the meeting held in the Washington, D.C. area. At that time, it was felt that ultrasonic propagation properties, mainly attenuation and the speed of sound, and these properties as functions of the acoustic and the state variables, could be used to characterize tissues according to functional, structural, and perhaps teleological criteria. The idea was that if one had detailed measurements of these properties, one could assign resolvably unique values to each tissue structure, and this then would allow differentiable values of the pathological state. Since that time, however, investigators learned just how difficult this problem really is. However, it seems that at the present time we have reached a stage where we have adequate technology to obtain these detailed empirical data, and so we expect to see a resurgence of interest in this area and substantial progress being made.

The papers here exhibit the richness of ideas and approaches that have been taken in this field. The symposium presentations were recorded and transcribed. The speakers then edited the transcriptions, and these comprise the proceedings. The discussions following each talk were also recorded, transcribed, and edited only for continuity and completeness. Thus the richness and spirit of the discussions have been preserved.

Professor Hill raises a very important question: can tissue characterization actually become a science? He discusses how to introduce objectivity as a measure of reliable performance. He takes as a possible axiom that an unknown volume of tissue is uniquely characterized by a particular set of values of underlying physical properties, with differences in such values corresponding to differences in tissue type and state. Thus the topic is seen as a sort of branch of signal communication science. Also, in a sense this is perhaps similar to the original ideas that were proposed or suggested approximately two decades ago.

The next paper, by Dr. Inoue, presents and discusses a model, essentially a biological model, for making pertinent measurements. The model comprises ground liver and ground fat to provide a known composition to be measured and to provide an understanding of how absorption and, possibly, velocity vary with tissue composition.

Dr. Waag discusses ultrasonic tissue characterization from scattering measurements as a function of angle and frequency to exhibit the contributions from compressibility variations and density variations, as well as the correlation of these two parameters. Together with modeling and measurements, his results seem to show that determinations of power spectra of the scattering medium variations can be made under practical conditions, implying the significance of density variation contributions to scattering. Much of his work has been done with calf liver. These analyses and measurements show that reliable, empirical methods now exist for tissue chracterization from ultrasonic scattering. The clinical usefulness of these methods for diagnosis may require further study.

Dr. Ohtsuki and his associates have approached the problem of determining the ultrasonic wave field in homogeneous media, that is, tissues, by the development of the ring function method. The method provides for computation of the entire field produced by an arbitrarily shaped finite source.

Dr. Hachiya and his associates discuss the speed of sound as a tissue-characterizing parameter in terms of a non-contact measurement method. This is very important, because contact methods have a way of distorting the tissue path.

Dr. Thijssen has devoted a great deal of attention to the question of tissue characterization and has shown that the effects of the transducer and other equipment performance characteristics really have to be understood and taken in account for successful tissue characterization. He shows that a truly multiparameter approach is required.

Dr. Ophir has established the relatively new field of elastography, which involves acquisition of RF ultrasonic scans before and after rapid compression of the tissue being investigated. Elasticity properties such as Young's modulus-type information can be obtained. This methodology is very promising and is expected to provide a significant method of characterizing tissues and their pathologies.

Dr. Kanai and his associates propose a new method for obtaining the local pulse wave velocity associated with an index of hardness of tissue. They are especially interested in applying this to the aorta. They have carried out model experiments which suggest strongly that the methods will be useful.

Dr. Miller has devoted a great deal of attention to quantifying the acoustic properties of heart muscle needed to assess function and optimal diagnosis. He has observed that the acoustic contrast responsible for echographic imaging arises from impedance differences between the extracellular collagen network that surrounds the myocite and the remaining tissue. He has proposed a model of the elementary scatterer as an ellipsoidal sheet having the properties of wet collagen embedded in a host material of properties of the average myocardium. This has been quite successful in describing the observed experimental results.

Because the morning session of the symposium ran late due to the involvement of many participants in the discussions, the remarks of Prof. Chubachi and Mr. Suganuma were necessarily abbreviated. Thus, Prof. M. Tanaka undertook to prepare, for the proceedings, a very brief history of ultrasonic microscopy and a description of the Tohoku University system. This presentation precedes the papers dealing with ultrasonic microscopy. The discussion that followed Mr.

Introductory Remarks xi

Suganuma's brief presentation is preserved and follows Dr. Tanaka's presentation.

Dr. Jones discusses clinical dermatology from an acoustical microscopic point of view.

Dr. Saijo further pursues ultrasonic microscopy tissue characterization, particularly of tumor tissues. He shows that these methods can provide mechanical, physical properties unobtainable by any optical methods or other microscopic methods. These kinds of data provide information for understanding details of echographic imaging. These are illustrated with studies of gastric and renal cancer material.

Dr. Tanaka discusses abnormalities of the myocardium in a quantitative fashion and illustrates the kind of useful information that is available from acoustic microscopy.

The topic of ultrasonic tissue characterization is addressed from many points of view, and it becomes clear where the field stands at present and where it must progress in order to make the techniques ever more useful.

F. Dunn, Ph.D.

Contents

Preface   vii

Introductory Remarks   ix

Authors   xvii

I   BASIC APPROACHES

1   Can Tissue Characterization Ever Become a Science?   3
- 1.1   Quantitative Ultrasonology   3
- 1.2   Precision and Relative Contrast   6
- 1.3   Spatial Resolution and the Quality of Information   7
- 1.4   Interrogation Rates   8
- 1.5   Multiple Parameter Confusion   8
- 1.6   Conclusions   9

2   Ultrasonic Absorption Measurement of a Model of Fatty Liver Tissue   19
- 2.1   Materials Studied and Method   20
- 2.2   Temperature Rise Measurement   23
- 2.3   Estimation of Absorption   25

3   Ultrasonic Characterization of Tissue from Measurements of Scattering as a Function of Angle and Frequency   31
- 3.1   Measurement of Scattering Properties   31
- 3.2   Application of the Model   34
- 3.3   Measurements of Calf Liver   37
- 3.4   Contributions of Compressibility and Density Variations to Scattering   39
- 3.5   Correlation of Scattering with Tissue Morphology   43
- 3.6   New Apparatus and Additional Investigations   46
- 3.7   Conclusion   47

4 Sound Field of Disk and Concave Circular Transducers — 53
4.1 Introduction — 53
4.2 Expression of Sound Pressure Emitted from a Transmitter with Ring Function — 53
4.3 Some Examples Using the Ring Function — 57
4.4 Examples of the Sound Field — 60
4.5 Conclusion — 62

5 Non-Contact Measurement of Sound Speed of Tissues — 63
5.1 The Measurement System — 64
5.2 Sample Preparations — 65
5.3 Measurement Results — 66
5.4 The Results at View Point of Fat Content — 69
5.5 Conclusion — 70

II CLINICAL APPLICATIONS

6 Multiparameter Ultrasonic Tissue Characterization and Image Processing: from Experiment to Clinical Application — 75
6.1 Introduction — 75
6.2 Parametric Imaging — 76
6.3 Image Processing — 83

7 Elastography: A Method for Imaging the Elastic Properties of Tissue in vivo. — 95
7.1 Previous Works on Tissue Elasticity — 96
7.2 Theory — 97
7.3 Elastography — 99
7.4 Basic Technology — 101
7.5 Results — 110
7.6 Conclusion — 118

8 High-Resolution Measurement of Pulse Wave Velocity for Evaluating Local Elasticity of Arterial Wall in Early-Stage Arteriosclerosis — 125
8.1 A Method for Measuring Small Vibrations — 126
8.2 Application to the Diagnosis of Arteriosclerosis — 126
8.3 Water Tank Experiments — 130
8.4 *In vitro* Experiments — 132
8.5 *In vivo* Experiments — 132
8.6 Conclusions — 136

Contents

9 Some Relationships between Echocardiography, Quantitative Ultrasonic Imaging, and Myocardial Tissue Characterization   **139**
- 9.1 Tissue Characterization in Insulin-dependent Diabetics · · · · · 146
- 9.2 Indirect Applications of Tissue Characterization · · · · · · · · · 149
- 9.3 Myocardial Anisotropy · 150
- 9.4 Summary: Current and Future Contributions of Tissue Characterization · 155

III   ACOUSTIC MICROSCOPY

10 Acoustic Microscope for the Tissue Characterization in Medicine and Biology   **171**
- 10.1 Introduction · 171
- 10.2 Acoustic Microscope System in the Frequency Range 100-200 MHz 173
- 10.3 Measurement Method of Acoustic Properties of Thin Tissue Specimens · 180
- 10.4 Quantitative Two-dimensional Display Method of Acoustic Properties · 190
- 10.5 Data Acquisition of Acoustic Properties of Biological Tissue · · · 193

11 Applications of Acoustical Microscopy in Dermatology   **201**
- 11.1 Equipment and Imaging Examples · · · · · · · · · · · · · · · · · 201
- 11.2 Acoustical Microscopy of Basal Cell Carcinoma · · · · · · · · · 206
- 11.3 Quantitative Methods and Conclusions · · · · · · · · · · · · · · 210

12 High Frequency Acoustic Properties of Tumor Tissue   **217**
- 12.1 Introduction · 217
- 12.2 Materials and Method · 217
- 12.3 Results - Gastric Cancer · 218
- 12.4 Discussion - Gastric Cancer · 223
- 12.5 Results - Renal Cell Carcinoma · · · · · · · · · · · · · · · · · · · 223
- 12.6 Discussion - Renal Cell Carcinoma · · · · · · · · · · · · · · · · · 228
- 12.7 Conclusions · 228

13 Acoustic Properties of the Fibrous Tissue in Myocardium and Detectability of the Fibrous Tissue by the Echo Method   **231**
- 13.1 Introduction · 231
- 13.2 Method and Materials · 232
- 13.3 Attenuation and Velocity of Ultrasound in Fibrotic Tissue in Myocardium for Cases of DCM, HCM, and CS · · · · · · · · · · · · 234
- 13.4 Shape and Structure of the Boundary Surface between Normal and Abnormal Tissues in Myocardium · · · · · · · · · · · · · · · 236
- 13.5 Relationships Between Changes in Quality of Abnormal Tissue and Intensity of Reflection or Reflectivity at the Abnormal Tissue Boundaries · 238

13.6 Detectability of Echoes at the Tissue Boundary by the Use of
     Commercially Available Echocardiographic Instruments · · · · · 241
13.7 Conclusion · 243

Closing Remarks 245

Index 247

Authors

C.R. Hill, Ph.D. Royal Marsden Hospital, Stoney Bridgy House, Castle Hill, Axminster, Devon, EX13 5RL, UNITED KINGDOM

H. Inoue, Ph.D. Department of Electrical Engineering, Akita University, 1-1 Tegata Gakuen Machi, Akita 010, JAPAN

R.C. Waag, Ph.D. Department of Electrical Engineering, University of Rochester, Medical Center, Rochester NY 14610, U.S.A.

S. Ohtsuki, Ph.D. Precision and Intelligence Laboratory, Tokyo Institute of Technology, Nagatsuta 4259, Midori-ku, Yokohama 226, JAPAN

H. Hachiya, Ph.D., Dr.MS. Department of Information and Computer Sciences, Chiba University, Inage-ku, Chiba 263, JAPAN

J.M. Thijssen, Ph.D. Department of Ophthalmology, University of Nijmegen, 6500 HB Nijmegen, THE NETHERLANDS

J. Ophir, I. Cespedes, N. Maklad, and H. Ponnekanti Departmet of Radiology, MBS 2130, University of Texas Medical School, 6431 Fannin Street, Room 6168, Houston TX 77030, U.S.A.

H. Kanai, Ph.D. Department of Electrical Engineering, Faculty of Engineering, Tohoku University, Sendai 980-77, JAPAN

J.G. Miller, Ph.D. Department of Physics and Cardiovascular Division, Washington University, St. Louis Missouri 63130, U.S.A.

M. Tanaka, M.D. Department of Medical Engineering and Cardiology, Institute of Development, Aging and Cancer, Tohoku University, Sendai 980, JAPAN

F. Dunn, Ph.D. Bioacoustics Research Laboratory and Computer Engineering, University of Illinois, 1460 West Green Street, Urbana, IL 61801, U.S.A.

N. Chubachi, Ph.D. Department of Electrical Engineering, Faculty of Engineering, Tohoku University, Sendai 980-77, JAPAN

R. Suganuma, B.S. HONDA Electronics co., Ltd., 20 Oyamazuka Ohiwa-cho, Toyohashi Aichi 441-31, JAPAN

K. Honda, Ph.D. HONDA Electronics co., Ltd., 20 Oyamazuka Ohiwa-cho, Toyohashi Aichi 441-31, JAPAN

J.P. Jones, Ph.D. Department of Radiological Sciences, University of California, Irvine, California 92717, U.S.A.

Y. Saijo, M.D., Ph.D. Department of Medical Engineering and Cardiology, Institute of Development, Aging and Cancer, Tohoku University, Sendai 980, JAPAN

H. Sasaki, M.D. Department of Medical Engineering and Cardiology, Institute of Development, Aging and Cancer, Tohoku University, Sendai 980, JAPAN

Part I
BASIC APPROACHES

Chapter 1

Can Tissue Characterization Ever Become a Science?

Professor Tanaka pointed out in his introduction that there has been a problem in ultrasonic tissue characterization , in that we have not been perhaps as quantitative as we should like. What I would like to do today is to try to open and stimulate some discussion on whether we can perhaps move toward a more quantitative approach to the questions of: why we are doing what we are doing? and, how well can we perform?.

1.1 Quantitative Ultrasonology

The topic is therefore: "How well can we perform in quantitative ultrasonology, bearing in mind in particular that ultrasonic tissue characterization is moving toward creating images that are in the form of the display of some parameter?"

In the European and in particular Mediterranean culture, there is a long tradition of building very tall buildings. Fig. 1.1 shows one from Greece. There is a very old traditional story in European culture of a place in what is now Palestine, which is called in Arabic "Babel", where a long time ago there was a plan to build a very tall building. The problem was that all the people who came there to build it spoke in different languages and did not communicate with each other very well. And this is the sort of problem that the world of ultrasonic tissue characterization has reached.

Figure 1.2 illustrates this. Ultrasonic tissue characterization typically has very many parameters and quantities that we try to work with. Attenuation coefficient, the power of frequency dependence of absorption coefficient, backscat-

---

*C.R. Hill, Ph.D.

Fig. 1.1 A building in Greece

tering coefficients, nonlinearity parameter, movement parameter, elastography parameters, and so forth. And we have many different people working on these different areas. Things have changed, but until quite recently we had only rather small data processing capacity on which all this could be based.

But, just along on the next hillside, there is another tower: the people from NMR and MRI who have only two or three parameters and a lot of processing ability. And what we in ultrasound have to do is to think about these different quantities–why we are using them, whether we use them all in some combination or just a few, and how we assess which are the good ones.

Just for those of you who are not so familiar with the field, some of the quantities that we try to use in characterizing tissues with ultrasound are listed in two groups as follows:

Class of Property/Characterizing Parameter(s)

- Bulk Properties

    - Attenuation Coefficient
    - Backscattering Cross Section
    - Speed of Sound
    - Non-linearity Parameter
    - Shear Elastic Modulus

- Spatial Organization

    - Frequency Spectra Features
    - B-scan Texture Features

How do we make quantitative comparisons between these? We should look at this in several different ways as measures of performance of tissue

1. Can TC Become a Science?

Fig. 1.2 Babel in ultrasonic tissue characterization

- Precision   (%)
    - local
    - universal

- Relative Contrast Index   (Nr)

- Spatial Resolution   (mm)

- Interrogation Rate   (Hz)

I shall deal with these in turn but, at this stage, you will appreciate of course that this scheme is not just limited to ultrasound. This is an approach that should be generally applicable to all kinds of modalities that attempt to characterize tissues. And you will see later on that in fact the NMR people have been working on similar lines.

My purpose in presenting this material today is to open the subject for discussion. For those who are interested, a fuller and more detailed treatment is available in published form (1).

1.2 Precision and Relative Contrast

Well, what about precision? What do we mean by precision? I would suggest that we should think of "precision in tissue characterization" in terms of the percentage measurement error that is experienced in repeating a nominally identical measurement of a given tissue parameter

(a) "locally": using any one particular equipment /technology.

(b) "universally": using any equpment /technology.

Even if you use one particular equipment, there will be some noise, thus leading to variations in repeated measured values of a given quantity. Furthermore, if you are trying, for example, to measure absorption coefficient of a tissue, the different equipment and different techniques will probably give you different answers.

So precision, presumably, should be the best possible, but "local precision" will be a component of what I'm going to talk about as the "relative contrast index". "Universal precision" the precision with which somebody in Sendai can measure things at the same time that somebody in London can measure things and communicate with each other–this will be the key to widespread use outside research centers.

So what about relative contrast index? What we have to think about here is that what we are trying to do in tissue characterization is to make some differential diagnosis, some distinction between one tissue and another. So "Relative Contrast Index for tissue characterization" should be possible to define in terms of what is "the number of categories of tissue type that can be distinguished using a particular equipment and technique at a given (e.g. 95%) level of confidence." And of course this would only have meaning in relation to a given diagnostic problem (e.g. in distinguishing between different kinds of diffuse liver disease) but it is the practically relevant quantity.

What should we aim at here? Well, certainly we must have at least a value of two: we must be able to distinguish at least between two different categories of tissue types. The important criterion is the confidence that it attributes to a given differential diagnosis. And there has been talk in the past that NMR is much better than ultrasound "because it is more physiological". I think this is a false argument. And part of what I want to propose is that we in the ultrasound community should have a basis for making comparisons with techniques like NMR so that we can assess how we are doing in comparison with NMR techniques. And therefore when a radiologist decides to solve a particular problem, he should have some objective approach to knowing which technique to use.

1. Can TC Become a Science?

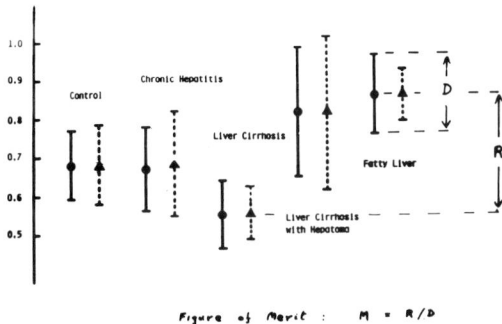

Fig. 1.3 Figure of merit

In talking about relative contrast index, one can define some sort of figure of merit, some sort of quantity that tells you how good a particular technique is for distinguishing between two different tissue types. In the paper that I have already referred to (1), you will see that there is some sort of simple mathematical derivation of figure of merit. In fact, having done this mathematical derivation, I discovered that the NMR people have really come to almost the same quantity. Figure 1.3 provides an illustration of the relationship between the relative contrast and the corresponding variances: the example given comparing the attenuation coefficient in liver cirrhosis and in fatty liver.

1.3 Spatial Resolution and the Quality of Information

What about spatial resolution? "Spatial Resolution " in tissue characterization is more straight forward: for a particular tissue characterization procedure, the linear separation (mm) between the center points of two pixels for which the tissue parameters can be separately measured. That is the sort of proposal that I would make.

So what should we be aiming at? In tissue characterization we probably cannot expect to be achieving spatial resolution that is quite as good as the best anatomical resolution that is obtainable in a B-scan. But very large regions of interest may become clinically meaningless. So, just as a sort of target to aim for, I would suggest that, in linear terms, we should be aiming to achieve something like "three times the best anatomical resolution" (i.e. a factor of 10 in area). And of course, we all realize that there is some sort of analog of Heisenberg's Principle for tissue characterization that there is generally a trade-off between the quality of information derived from an area and the spatial size of that area. In other words, the more tissue you sample the better quality of information you can get.

Here is an example for which I did a simple calculation, of what is the effect of the linear size, the L in millimeters, of the region of interest on the variance σ^2 of estimates of attenuation parameter S derived from a zero-crossing method.

Attenuation of Phantom

L(mm)	S(kHz/mm)	σ^2(%)
5	8.4	95
45	6.2	2
110	5.4	7

(Measurements at 3MHz and ROI width 60mm)

You can see that, for a small linear size, the variance is very great. For a large linear size, the variance becomes quite small, but this is not really a very useful size for which to do tissue characterization.

1.4 Interrogation Rates

What about interrogation rates ? I would suggest that this is the frequency (Hz) with which a complete set of parameter pixels (e.g. constituting a parametric image) can be measured and presented /displayed.

What should we be aiming at? If possible, we all know that we like to work in real time ($\sim$ 10Hz). Anything slower may not get used routinely unless it is very powerful indeed. But if we, again, make comparisons with our colleagues in NMR, real time can be $\sim 10^{-3}$Hz, Maybe they are getting a little bit faster, but NMR is really quite a slow procedure. So ultrasound here has a great advantage. And we are starting to see this now that we are getting the processing power that enables us to carry out tissue characterization work in real time.

1.5 Multiple Parameter Confusion

My final topic is what I have called multiple parameter confusion . This is an important thing to consider. If, for example, you try to measure the backscattering coefficient of a tissue, you can only do this in a truly quantitative way if you know the precise values of the overlying attenuation coefficient. It is absence of such knowledge that leads to so called multiple parameter confusion. It is the uncertainty or confusion that exists when quantitative imaging is attempted on the basis of one parameter which is not fully separable from another parameter, that cannot itself be independently evaluated. Again, making comparison with NMR, the NMR people try to measure T1, and they talk about T1- weighted images. They too have a problem in that they can never quite successfully separate one parameter from another.

1. Can TC Become a Science?

I just want to take as an example of this some very nice Japanese work that is trying to measure the nonlinearity parameter, B/A, and creating images from this.

There is a basic equation on B/A:

$$B/A = KP_2(x) \, e^{2\alpha x} - 2 \qquad (1.1)$$

The basic equation includes a term with an attenuation coefficient α. The range of values of nonlinearity parameters B/A in tissues is in the region of between 6.0 and 8.0. And you will find that this range will be obscured if you have even a 2 percent uncertainty in the value of attenuation coefficient α. So, in planning in what particular directions we put our energies in tissue characterization, we should think carefully about whether this sort of problem is going to arise and make clinical implementation not very practical.

And here is another example–the effect of a 10 percent uncertainty in attenuation of overlying tissues on estimates of bulk backscattering cross-section.

Attenuation of Tissues

Tissues	True Value	Uncertainty Limits
	$(cm^{-1} Sr^{-1} \times 10^{-4})$	
Human liver	2.2	0.84 ~ 5.7
Liver tumour	1.0	0.4 ~ 2.6
Beef muscle	9.2	3.5 ~ 24

("true values" Nassiri and Hill, 1986, at 4 MHz; Reference 2.)

The uncertainty limits are really very great. So, measuring bulk backscattering cross-section can be problematic if you want to achieve absolute values. Of course, that is not always necessary in order to do a good job of clinical diagnosis.

1.6 Conclusions

In the work we need to do on tissue characterization, there is a very exciting future for younger people who are joining the field just now. We certainly want to do a lot of work on fundamental studies on the tissue physics, but we surely want to develop methods that are going to be clinically practical. And also they will only be practical on a large scale if they are commercially practical. And this means that they have got to meet some sort of performance criteria. I have suggested some kind of approach to achieving performance criteria. These particular suggestions may well not prove to be optimum, but what I hope you all agree on is that some kind of performance criteria must be set up and achieved if we are going to get machines that manufacturers can make, and that can be used worldwide with confidence, so that different people in different clinical centers can understand the significance of the work of other people.

And so I would suggest that these performance criteria should be something like the ones we have just been discussing. Precision, which means repeatability. Relative contrast in the context of a particular problem. Spatial resolution, and speed and if we have not achieved "real time", we are going to have "hard times"!

I have suggested that you could roughly categorize tissue characterization procedures in two groups: into bulk tissue properties and parameters of tissue organization or structure. A lot of these are going to be difficult. Some, such as acoustic shear wave velocity–elastography, as some people call it–could be very promising. But some others are going to be doubtful. The ones that I would put my money on are going to be parameters to do with tissue organization, because they have got a prospect of meeting good performance criteria.

The condition for our success will be general acceptance for some sort of optimum technical approach which will bring with it associated commercial availability and standardization. And remember always that these techniques will never succeed unless they are sufficiently widely accepted, to the point that there can be a clinical learning program in many different centers. And this again underlines the need for some approach to standardizing the whole situation.

I have thrown out a number of proposals which a lot of you will want to criticize, and I hope, indeed, that this can develop into a continuing discussion, even after this meeting.

References

1. Hill CR, Bamber J C, Cosgrove D O(1990) Performance criteria for quantitative ultrasonology and image parameterisation. Clin. Phys. Physiol. Meas., (Suppl. A): 57-73.

2. Nassiri D K, Hill C R (1986) The differential and total bulk acoustic scattering cross sections of some human and animal tissues. J. acoust. Soc. Amer. 79: 2034-2047.

Discussion

DUNN: Thank you, Dr. Hill. Dr. Hill has been particularly provocative. Just to show you how provocative he was, when he chose to speak about the nonlinearity parameter and how it may not be a very successful tissue character parameter, he chose the method, to discuss which is least likely to be used. And I think he's been equally provocative in other areas of his talk, and this was purposeful on his part, because he wanted to provide grounds for discussion. And now we have that opportunity for discussion. As regards the nonlinearity parameter measurement, there are three methods available, and the one that is most likely to be used clinically is a different method than the one Dr. Hill discussed. Dr. Ophir.

OPHIR: I have a more general question regarding the title of your talk. I think the title implies that tissue characterization today is not a science. You really did not address this, and I would be interested in your definition of what a science really is and why you feel that tissue characterization does not meet that definition.

HILL: The point I was trying to make is that different people are working in different labs, and very often their results are difficult to compare. Perhaps the title was too provocative, and perhaps I should have put the emphasis more on taking the science into a practical technology. But I certainly didn't mean to offend a lot of very excellent scientists like yourself, who I know are doing very good science.

DUNN: I'm a bit uncertain, but Dr. Hill may have invented a new term about 10 or 15 years ago; telehistology. I'm not sure whether you invented it, but you certainly have been using it. By this term, he essentially meant that tissue characterization could be, in some way, made analogous to the field of histology, which for many people over the years has been thought to be not a science but a black art. And I wonder if Dr. Hill had something like that in mind when he initially proposed the term. Is it more like a black art?

HILL: Well, not at all. Histology literally means the science of tissues, the study of tissues. It's a Greek word, as you know. And telehistology is, by analogy with television, and so forth, viz., to do histology by remote means. And maybe histology is to some of us a bit of a black art, but I think histologists would claim that it is a science, and I used telehistology really because it seemed to be a term that had attractive parallels with the words we use for other remote procedures. I think it puts emphasis on the study of tissues, and brings a sort of direct comparison to the user with the techniques that they know in sending tissue specimens to the pathology lab.

THIJSSEN: You mentioned the uncertainty principle that was invented by Heisenberg, but I think physics is much easier than tissue characterization. So we have a lot of extra uncertainties involved. And I would like to go back to the first part of the lecture where you introduced the local and universal precision in the measurements. I think that in addition to the aspects you mentioned, we should add at least that in addition to precision, we also have the accuracy, or bias, of a method that can be present in our results. And secondly, I would like to distinguish between the effects of the measurement technique we are using as compared to the intrinsic biological variability. And that's the extra problem we always have when we measure biological effects or parameters. Hopefully you can explain again what you meant, i.e., because to me it was not that clear the difference between local and universal and the relation to technique versus intrinsic variability.

HILL: Well, thank you for bringing that point out. What I should have said, although I did say this in the abstract, is that what I've been talking about is not new, obviously. In fact, it is based on a paper that some colleagues and I published three years ago[1]. That deals with the point that Dr. Thijssen made in more detail. What I'm trying to bring to people's mind is that there are questions of precision and there are questions of accuracy. Precision and accuracy are two different things. Accuracy is how well you can measure a quantity in absolute terms in relation to what its true value is. Precision is how well you can make a repeat measurement of the same quantity, even though you do not get the value to its absolute and correct value. In thinking about precision in tissue characterization, we have to think that there will be variations of some particular tissue property within any given patient. The absorption coefficient of normal human liver varies probably between different parts of the same liver. It will certainly vary between different individuals. That's part of normal biological variability. So that's one factor of the noise that is present. We then find that even if we use the same machine to make these measurements, there will be some noise, some variability brought in just because the machine varies from day to day, etc. So these are two components of local variability. And then the thing becomes more complex because you use another machine of the same type that's a bit different. You use another machine of a different type, and that's different again. So really what I'm trying to set out is the thought–and I don't think that we've done this sufficiently systematically in the past–that you can gradually make an approach to thinking about what are the errors potentially in what we're trying to do? And how do we take account of these, and how do we report them, and how do other people know how important these errors are in comparing our results with their results? Does that answer your question?

THIJSSEN: More or less. I will just make one more remark, and maybe you can answer that. But I don't think it's a question. The point is that part of the variability difference has been a problem in the past between different machines and different institutions. We have tried to solve this, at least in Europe. We have worked some years in some cooperative efforts to find corrective strategies to eliminate equipment performance in the measurements. I think on one of your slides there was a prerequisite for acceptance of any tissue characterization method that it should be equipment, institution, and researcher independent. I think we have achieved that to a large extent. I hope people are aware that it's possible to get rid of these problems.

JONES: Just two comments. One is related to acoustical modeling, which Professor Hill suggested in his presentation. He gave what I thought was a very nice example of measurements of the nonlinear parameter, B/A, and that the investigator must be careful in doing such measurements because they're influenced by the attenuation values. This leads to the comment that any acoustical measurement that we make, any image that we make using ultrasound, is based on some tacit assumption of the acoustical model. If the model is incorrect, and most of the models we use today are certainly only approximations of that interaction process, we are apt to produce images that may have very high resolution, but in fact may have very little information content. And so this is a problem in tissue characterization that we need to address to make sure that the basic models that we have on the interaction of tissue with the ultrasound are correct. Or, at least, a better approximation, perhaps, than we have now.

The second comment is related to other imaging modalities. Those of us in Departments of Radiology, deal daily with MRI and comments from radiologists who complain greatly about the rather fuzzy ultrasound images. I think it's worth noting, and Professor Hill certainly hinted at this, that the interaction of ultrasound with tissue is a very complicated and complex process, and we have a very difficult problem. It's a very challenging problem, and I'm very pleased to see the number of younger people here who are entering the field. I think it's a very challenging field and one that is going to have a lot of opportunities, but we clearly have a long way yet to go and many challenging problems to work on.

WAAG: I would just like to ask Dr. Hill to clarify one remark he made and perhaps ask for a response from Dr. Ophir. I thought I heard, at the end of your presentation, that you classified elastography in the group of techniques that looks at bulk properties of tissue. And then you suggested that some pessimism existed in your mind about the success of that. I would like you to say whether that's a correct paraphrase of your remarks, and then I would like to have Dr. Ophir reply.

HILL: Well, I get into trouble with my own colleagues. I perhaps should have said that the reason I wrote this paper in the first place was that it came out of discussions we had in our own group as to why we were doing what we were doing. And my own colleague, particularly Jeff Bamber, disagrees with me on this classification. So I quite accept that this is not altogether a satisfactory way of classifying things. Among the "bulk" properties, shear modulus, I think, is noticeably a different one. What I do feel, though, is that some of the quantities like absorption coefficient and backscattering may not be the most promising ones, in the long term, to make use of in a general way. There will always be exceptions to this and there will be some examples where this is just exactly the right thing to do. Shear modulus in some ways, deals with the entire way that tissues are structured. That was indeed why people like John Ophir and ourselves got interested in the first instance, because tumors particularly we know have got a different structure and different elastic properties. I'm not sure that I've answered your comment sufficiently well. Perhaps I'm saying that I wouldn't stand too closely to the scientists' categorization scheme that I've put forward. But in some ways and for some purposes I think it can be helpful.

DUNN: Thank you very much. Dr. Ophir, I'd like to hear from you on this.

OPHIR: I tend not to make this distinction between bulk and–I don't know what the other word is–local(discrete). I think it's a matter of degree, a matter of scale, as in acoustics, where everything scales with the wavelength. If the wavelength is small bulk becomes a more local entity. I think it's a continuum. I believe it's a matter of the instrumentation you use, the frequencies you use, and your abilities to resolve appropriate quantities. In elastography, for example, we know that from a fundamental point of view the ability to measure strains in tissues can be done on very small areas to a very high degree. So I don't know if I should call it bulk or local; that's important. I work in a department of radiology, and they always want to see resolution. They always want to see small things with high contrast. If I can perform and approach the theoretical limits of a given parameter, that would be fine. Even attenuation, which we all know from the experience of the last 15 years during which many of us were trying to estimate attenuation and it turned out that, if you want to do it in pulse echo mode, there's a lot of noise in the process. Therefore, you need large areas in order to make that assessment. When you go to large areas now, you say, well, this is a bulk kind of measurement that I'm giving you. But if we figure out how to reduce the noise and get to more sophisticated techniques, then maybe we'll be able to make attenuation measurements down to the submillimeter regions. The distinction, in my mind at least, is not very important.

TANAKA: Let me speak in Japanese. Dr. Hill, you have told us about a lot of parameters, and I thank you for that. I would like to hear your comment on the following. I think there is another important aspect to tissue characterization, and that is the distinction between static and dynamic characteristics. I think these are also important aspects. This is not an in vitro or in vivo issue, but within the living state, there is a dynamic situation as well as a static condition. I would say that the conditions differ among the two. As far as cardiologists are concerned, particularly those of us who are focusing on the myocardium, and therefore on both the static living condition and also on the dynamic, pulsating situation, I would say that acoustic tissue characterization will be quite different. From your parameters, Dr. Hill, how should we deal with this? If you can comment on this I'll be very pleased.

HILL: Well, I think you've made a very important and valuable point. I'm not sure how I can comment on this beyond making a rather obvious comment. The work that Dr. Ophir has just been referring to that he and others are doing on the elastic properties of tissue, which reflect the, as it were, dynamic properties of tissues, seems to provide a very exciting prospect for the future. I think what is surely happening is that it's only just now that we're getting sufficient computing power to enable us to make good sense of the data that ultrasound provides. Perhaps all I can do is agree with you in saying that if you compare ultrasound with other imaging techniques, one of the great strengths that it may show in the future is being able to go beyond measurement of static properties to measurement of dynamic, kinetic properties. That's a rather vague response. I can't make a better comment, but there are other people here who I'm sure could comment better than I. Thank you.

1. Can TC Become a Science?

DUNN: Is there anyone who wants to comment? Dr. Ophir?

OPHIR: Yes, I just wanted to state that there are others, as has been noted, that are working, for example, on elastic modulus determination in tissue. Notably, several groups are working on what is known as sonoelasticity images. In that area, they are indeed trying, if I understand what they are doing correctly, to characterize the dynamic moduli of tissue which might be different than the static moduli and delimit the dynamics at zero Hertz. So I think that the wealth of information that is available regarding elastic moduli may be indeed very high, and the results that you get will depend on the frequency that you use, which is really not surprising. However, the methods to be used might be quite different.

DUNN: Any further discussion of Professor Hill's paper?

HILL: Could I just say that I do have some copies of the paper on which my talk was based, and I'll leave these on the table outside if anybody is interested.

DUNN: I urge you to see this paper. It's a very interesting one.

NITTA: For the sake of the Japanese participants, I'll speak in Japanese. This symposium is very important to carry out detailed discussions on this, so please feel free to raise any comment or question.

QUESTION: I am an engineer and I am interested in instrumentation. I would like to ask Dr. Hill: Is there any instrumentation that you consider ideal, that you would like to have, or that you envisage as desirable? It may be difficult to specify immediately, but in your presentation you discussed instrumentation based on your ideas. So if you have any ideal instrumentation, or future instrumentation, that you have in mind, do you think that you can provide suggestions for the engineers in the room?

HILL: Well, I don't think I can give you a design of the ideal instrument. We all know that the procedure of pulse-echo interrogation is very powerful. Something that we're now seeing is the use of multiple element interrogation in parallel. Until now we have simply been receiving on one transducer, the echoes resulting from the propagation of the pulse, usually from that same transducer. We are going to be able, in the future, to receive simultaneously from a number of transducers, and this immediately increases our ability to derive important information about relationships between different tissue properties and different regions of the organ being studied. And from then on, surely a lot of the successful techniques are going to depend on the particular processing of image information that we carry out. Again, I'm not making a very fundamentally creative comment on your question, but I cannot think of anything better to say. But again, I'm sure that there are other people here who could comment much better than I can. I hope this discussion is not too much based on what I've been trying to say; I hope it can be a more general discussion.

OPHIR: I'd like to tease the audience a little bit more and maybe reinterpret my interpretations of Dr. Hill's comments. What I'd like to pose to the audience is a fundamental question of tissue characterization, and this is, does tissue characterization need to be quantitative? I think the assumption of Dr. Hill this morning, and of many of us, always has been that you must give numbers. Now that I think about the things he said, maybe I misunderstood his definition of bulk versus local. Perhaps, and Professor Hill can correct me if I'm wrong, maybe what he meant was that bulk is giving numbers and local means looking at the spatial arrangement of things that may be not very well quantified in terms of those numbers. I think there are many examples which you may, or may not, want to call tissue characterization that are popping up all the time, such as color Doppler, for example, which has been very successful in the last five or six years, and we all know it is not quantitative (there's a Doppler angle that we don't know about). In elastography, for example, we're going to give you tissue strain, which has to do with Young's modulus, but it also has to do with boundary conditions and the amount you push, and so on. I think there are many examples. Now you may not wish to call that tissue characterization, but is it permissible to broaden the definition of tissue characterization to the display of new kinds of images that, which in themselves are not quantitative, the spatial distribution of this new parameter, is in fact very useful.

DUNN: Does anybody wish to respond to the question, does tissue characterization need to be quantitative? When I referred to Dr. Hill's telehistology in comparison to earlier optical histology, of course this is a very broad, well-used field that has almost never been quantitative.

THIJSSEN: I would relate this question, and maybe also the answer, to the point that it depends on your goal. There are more examples than just color Doppler. We have now tissue Doppler imaging and things like that. I think it's correct to say that you need to have some numbers behind these images, but the matter is how precise you want to have them, and this depends on the goal. If you just want to see an image, you can relate the quantity imaged to the anatomy, and not in an absolute sense to figure out whether it's tumor A or tumor B. That's the first point. The second point is that you can also relate it to the question of the uncertainty principle we discussed before. The more precise you want it, the less resolution you will get. So it's a trade-off all the time, and I think it's all a gradual scale between having an exact number or exact as possible, and an image. And that is the dilemma. I think it will be solved by industry, which will hopefully find enough inspiration in the work we're doing to implement some of the techniques and present them by way of images that are attractive to the medical field.

HILL: Could I just say in a lighthearted way that this all depends on whether you're a physicist or an engineer. If you're a physicist, you really feel insecure if you don't have good numbers. If you're an engineer, you just want to make something that works. And of course, the engineers really win out, because you go to a meeting like some of us have just attended, the World Federation for Ultrasound in Medicine in Sapporo, and 90 percent to 95 percent of the work that is described in clinical applications really succeeds not because it is quantitative, but because it is practically useful; because the engineers have done a good job. I was trained as a physicist. I still think I would like a world which, in principle, we could make good measurements of real numbers. But I accept that the real world is a world of engineers and radiologists.

TANAKA: There was a mention about whether quantification is necessary for

1. Can TC Become a Science?

tissue characterization. We are conducting tissue characterization, and I think from the viewpoint of medicine, there two factors involved here in quantification. One is the measure of the absolute value; this is one part of quantification. And the other is the relative value. The determination of the quantitative value when we use organisms is very difficult, I'm sure that you know this, particularly because of the variation encountered in the body and the organisms. In clinical medical applications, I'd say that relative quantification of values is important, because abnormalities are compared to normalities, etc. The comparison between the normal and the abnormal in terms of quantity is, I think, in medicine most important. In terms of quantification, such relative measurements would be very important, particularly in cases of pathological tissues. Obviously we need to do the normal and the abnormal tissues simultaneously. Such observations will be required, I think, in the future. The quantification of tissue characterization will involve many difficult issues and problems. What do you say to this?

OPHIR: I think I agree. The only comment I have regarding the relative business is, do you mean relative within the same patient, or do you mean relative from one patient to the other? I think if you have relative within the patient, then you can see, for instance, the presence of a tumor on a background of normal tissue, and if the attenuation is a little higher or lower then it's immediately obvious. If you do it amongst patients, then maybe it's a more difficult problem.

TANAKA: Well, I'm talking about both–interindividual comparison and intraindividual comparison between normal and abnormal tissues. So there is always a comparison, but always between the pathological tissue and the normal tissue. That's what I'm interested in doing.

DUNN: As you all know, we have an extremely busy schedule today, and I think I would like to use the chairman's prerogative to move on to the next paper.

Chapter 2

Ultrasonic Absorption Measurement of a Model of Fatty Liver Tissue

The principal properties involved in ultrasonic tissue characterization are the speed with which the wave propagates and the absorption of the wave energy. Absorption is a material parameter, as it does not include the scattering and reflection properties included in the attenuation parameter. It is important that we obtain ever more absorption data of bio-materials in order to have a complete understanding of the interaction of ultrasound and living materials. Many studies of the ultrasonic analysis of tissues have been reported [1]-[11]. Here, analysis means that we have a material of unknown properties, and we wish to determine its characteristics. In the field of electrical engineering, for example, we also consider studies of "synthesis" of materials. First, analysis studies may be conducted, which lead to synthesis studies. The continuation of the processes leads to a deeper understanding of the media structure and to superior models. Thus, we think in terms of the synthesis of a model which has properties similar to that of human tissues, from which it may be expected that research on the origin of ultrasound absorption will be improved.

For this work, we chose to use mammalian liver tissue, as many other studies on analysis have been reported [2],[4],[7],[12]. Our goal is to create a model of fatty liver tissue which will be useful to clinical applications. The materials used in this study are pig liver and pig fat.

The two important properties to be measured are the velocity and absorption. The absorption data are presented here. Unfortunately, we could not determine the temperature characteristics of the absorption coefficient with high precision. The temperature rise produced by ultrasound irradiation will be shown first and then comparison with the temperature rise produced in 3 percent agar models

---
*H. Inoue, Ph.D.

Fig. 2.1 Microscopic image (ground pig liver x67)

will be made.

2.1 Materials Studied and Method

The materials studied, viz., pig liver and pig fat, were obtained from the market. The pig liver and pig fat were separately ground, in a food preparation blender, for several minutes and then mixed in various fractions. These specimens were then degassed. Figure 2.1 is a micrograph of ground pig liver. The scale unit on the right side of the figure is 0.1 mm. It is seen that the size of a ground liver grain is less than this scale unit. Figure 2.2 shows ground pig liver mixed with 25 % fat. Some ground fat remains large in grain size. The sample material was then wrapped with polyvinylchloride film, less than 10 μm thickness, for measurement. The measurements were performed at the room temperature.

Figure 2.3 is a block diagram of the measurement system [6],[11],[13]. A concave transducer, having a resonance frequency of 1.0 MHz, and a specimen supporting structure specially made for this work are mounted in the water tank. A thermocouple is inserted near the surface of the specimen and the temperature rise is detected using a preamplifier. The size of the thermocouples employed is of the order of 25 to 50 μm.

Figure 2.4 shows the equi-intensity ultrasonic field produced by the transducer. Figure 2.5 also shows the amplitude distribution of the acoustic field. Although the line of the axis of measurement is not perpendicular to the surface of the transducer, the specimen can be properly positioned at the focal point, at approximately 60 mm from the transducer.

2. Absorption Measurement of Fatty Liver

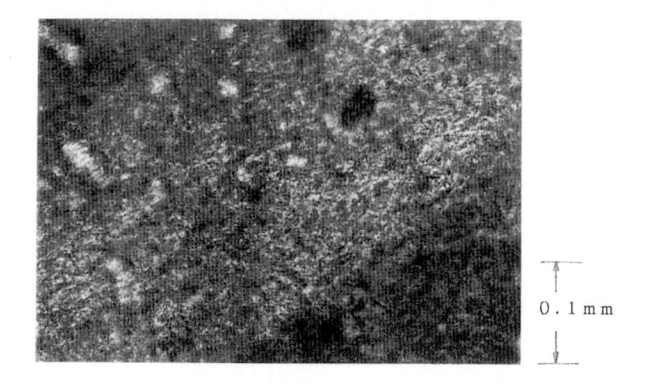

Fig. 2.2 Pig liver mixed 25% fat (x67)

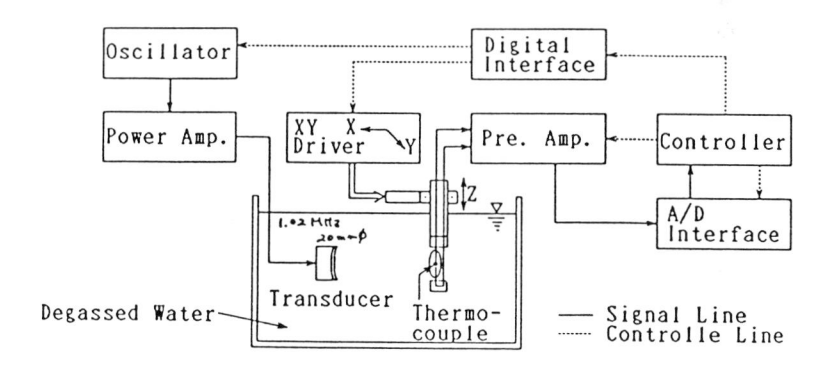

Fig. 2.3 Block diagram of the system

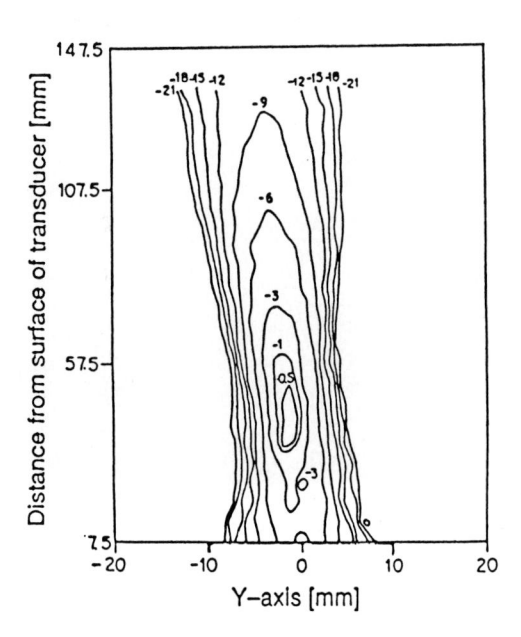

Fig. 2.4 Equi-acoustic intensity field of the concave transducer (20mmϕ, the radius of curvature 100mm, PZT) measured by a PVDF hydrophone (0.8mmϕ)

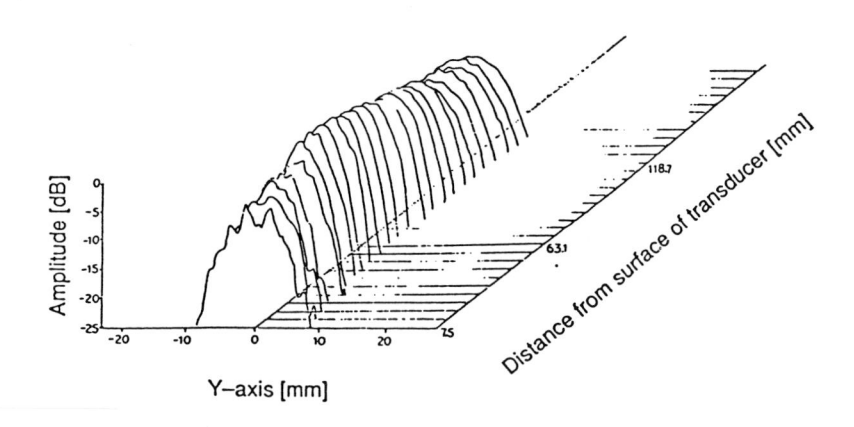

Fig. 2.5 Amplitude distribution of the acoustic field of the concave transducer (20mmϕ, the radius of curvature 100mm, PZT) measured by a PVDF hydrophone (0.8mmϕ) in three dimensions

2. Absorption Measurement of Fatty Liver 23

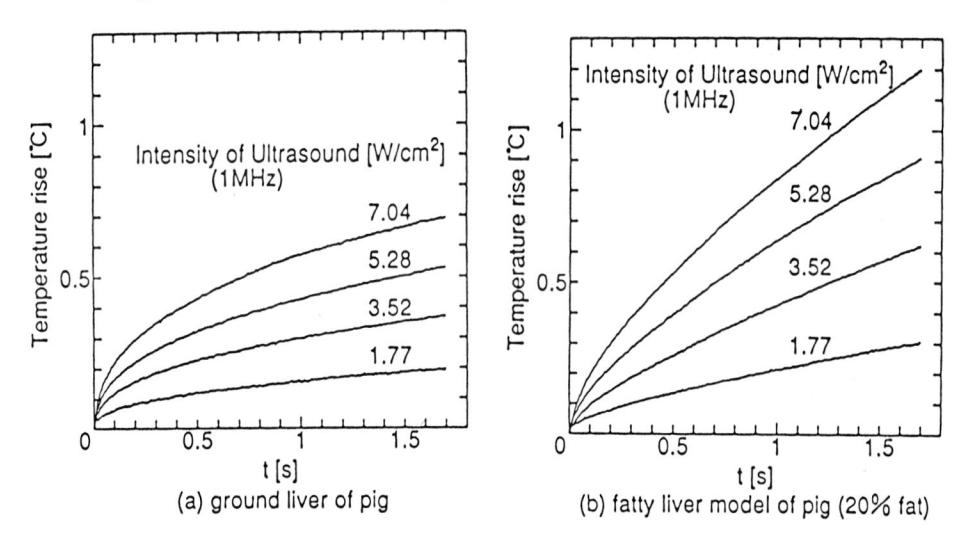

Fig. 2.6 Temperature rise produced by ultrasound irradiation

2.2 Temperature Rise Measurement

Figure 2.6 shows typical examples of the temperature rise produced by ultrasound irradiation, in the specimens, using a 25 μm diameter thermocouple. Four ultrasound intensities, as shown in the figure, were used. The maximum value of the ultrasound at the spatial-peak temporal-average power is 7 W/cm^2. The irradiation exposure time was 1.7 seconds. Figure 2.6(a) shows the temperature rise in pure ground liver. Figure 2.6(b) shows the temperature rise of a fatty liver specimen having 20% fat. The temperature rise in the fatty liver specimen is greater than that in the pure liver because of the greater absorption. From these data, we obtain the data shown in the figure 2.7. The abscissa is ultrasound intensity, and the ordinate is the observed temperature rise. The plotted points show the difference of the sample materials. The straight lines indicate the linear dependence of the temperature rise on the intensity. The curves of the solid fat and ground fat in the figure show the saturation characteristics of the temperature rise. Though it appears that pure fat exhibits a nonlinear dependence of absorption on wave amplitude, the origin of this non-linearity cannot be determined from these data alone.

Figure 2.8 shows the same kind of data obtained using a 50 μm thermocouple. As the smaller thermocouple results in a more stable temperature rise, we use the data by obtained with th 25 μm thermocouple for the following estimates [13],[14],[15].

Figure 2.9 shows the relative temperature rise in the materials, compared with a 3% agar preparation. The abscissa is the weight ratio of fat-mixed liver, and the ordinate is the temperature rise in the sample material relative to the temperature rise in the 3% agar preparation. 0% and 100% mean pure liver and

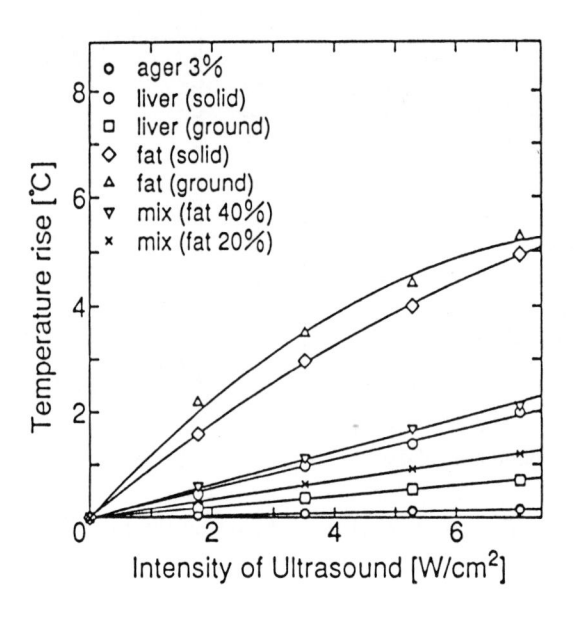

Fig. 2.7 Temperature rise after 1.7 s irradiation (25μmϕ thermocouple)

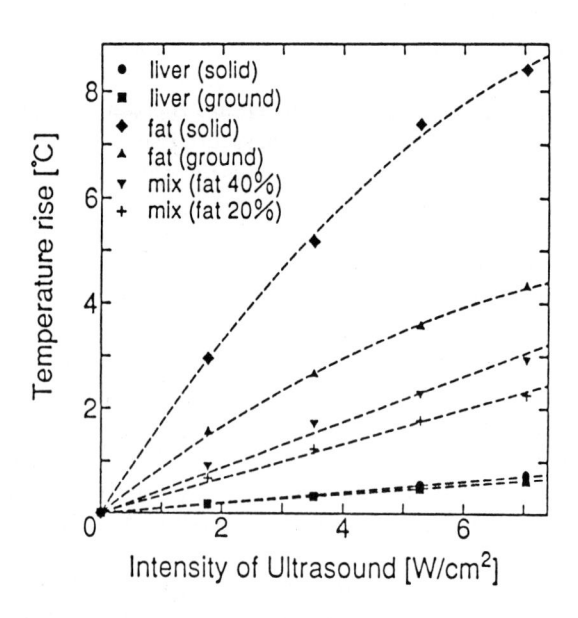

Fig. 2.8 Temperature rise after 1.7 s irradiation (50μmϕ thermocouple)

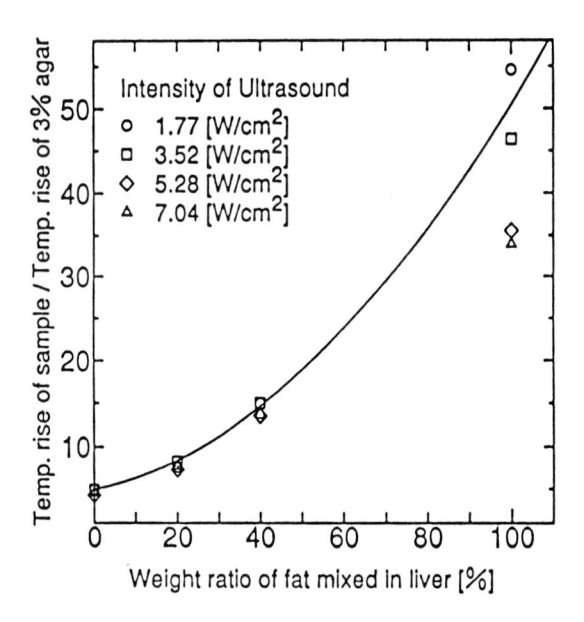

Fig. 2.9 Relative Temperature rise

pure fat, respectively. Samples of 20% and 40% of fat weight percentages were also measured. To avoid the nonlinear effects, the intensity chosen for estimating the absorption coefficient is chosen to be 1.77 W/cm^2. The curve drawn in figure 2.9 is a parabolic fit having a 0.95 correlation factor, viz., 2.56 x 10^{-3}x^2 + 1.20 x 10^{-1}x + 4.92 where x is the percent weight ratio of fat in liver.

2.3 Estimation of Absorption

Using the above data, we can estimate the the value of the absorption coefficient [11]. We employed a simplified equation from Parker [15],[16]

$$\alpha = \frac{\rho C(1 + (4kt/\beta))}{2I} \frac{dT}{dt} \qquad (2.1)$$

where β is the beam width of transducer, ρ is the density, C is the specific heat, k is thermal diffusibility, and $\frac{dT}{dt}$ is the time rate of change of the temperature rise.

This equation involves several assumptions in estimating the absorption coefficient. The measured quantities are the beam width, β, and the rate of temperature rise, $\frac{dT}{dt}$. We did not measure the other values, viz., density, specific heat, and diffusibility, which were obtained from Goss et al. [1],[14]. Figure 2.10 (a) shows the estimated absorption value versus time; four curves of estimated values for four different ultrasonic intensities. All of the estimated values agree

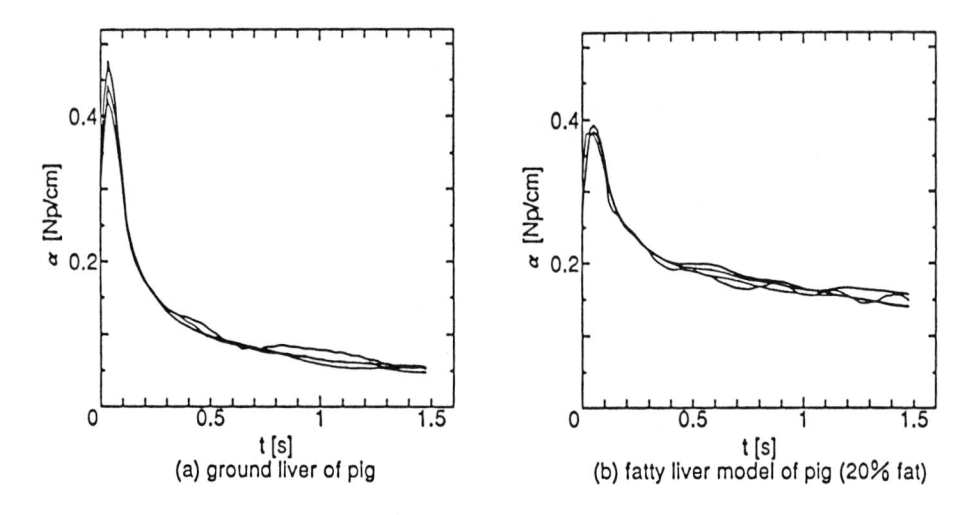

Fig. 2.10 Estimation of absorption coefficient α

because of the linear relationship of absorption and irradiation power. At the onset of the irradiation, there is a small temperature rise due to viscous action of the fluid-like tissue in contact with the metal thermocouple wires. A short-time later, thermal diffusion occurs. The absorption coefficient value at around 0.3 or 0.5 sec. is nearly the same as that earlier reported [2]. Figure 2.10 (b) shows the absorption coefficient of the 20% fat model. Superposing the figure (b) on figure (a) shows the difference of the absorption coefficient of the pure and fatty liver.

The velocity data for these same materials is included in fig. 2.11 [17]. The ordinate is the sound velocity and the abscissa is the weight percent of fat mixed in liver. The measurements were done at 37°C. The velocity of the pure liver is seen to be 1,600 m/s, and that of 40% fatty liver model is 1,510 m/s. A straight line can be drawn in the figure. As shown before, the absorption is not linear.

We have made a model of fatty liver, using mixtures of ground liver tissue with ground fat. The measurements of the sound velocity and absorption should be done in the same preparations for meaningful discussions of tissue characterization. We have shown here typical measurements of the ultrasonic properties velocity and absorption in a synthesized model. Such a synthetic model may be useful generally to investigate ultrasonic tissue characteristics of biological media.

Acknowledgment

I am grateful to Prof.Floyd Dunn, University of Illinois, for his advice and his help in the calibration of the transducer, and to Mr.Yasuo Yoshida, Miss. Maki

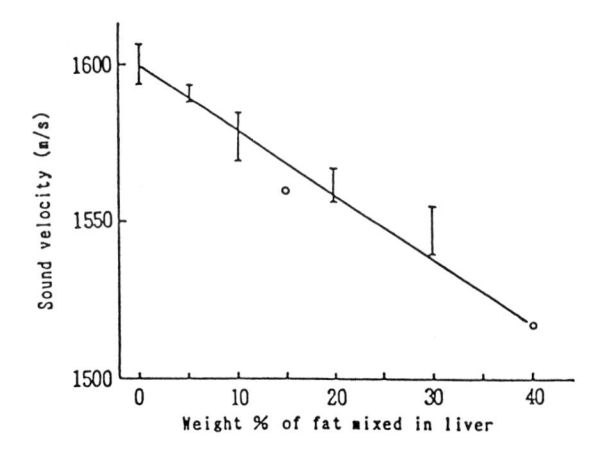

Fig. 2.11 Sound velocity vs. weight % of fat at 37°C(fatty liver model by pig)

Ishigo-oka, Mr.Hiroki Kobori and Mr. Kazuo Kato, Akita University, for their contributions to these experiments.

References

1. Goss SA, Johnston RL, Dunn F(1978)Comprehensive compilation of empirical ultrasonic properties of mammalian tissues. J. Acoust. Soc. Am. 64(2): 423-457

2. Bamber JC, Hill CR, King JA(1980)Acoustic properties of normal and cancerous human liver -II dependence on tissue structure. Ultrasound in Med. and Bio. 7: 135-144

3. Carstensen EL, McKay ND, Delecki D(1982)Absorption of finite amplitude ultrasound in tissues. Acustica 51: 116-123

4. Parker KJ(1983)Ultrasonic attenuation and absorption in liver tissue. Ultrasound in Med. & Bio. 9(4): 363-369

5. Ohkawai H, Nitta S, Tanaka M, Dunn F(1983)$InVivo$ measurement of thickness or speed of sound in biological tissue structures. IEEE Trans. on Sonic and Ultra. SU-30(4): 231-237

6. Carnes KI, Dunn F(1988)Absorption of ultrasound by mammalian overies. J. Acoust. Soc. Am. 84(1): 434-437

7. Parker KJ, Tuthill TA, Baggs RB(1988)The role of glycogen and phosphate in ultrasonic attenuation of liver. J. Acoust. Soc. Am. 83(1): 374-378

8. Parker KJ, Lyon MR(1988)Absorption and attenuation in soft tissues: 1-Calibration and error analyses. IEEE Trans. on Ultra. Ferro. and Freq. Control 35(2): 242-252

9. Lyon MR, Parker KJ(1988)Absorption and attenuation in soft tissues: 2-Experimental results. IEEE Trans. on Ultra. Ferro. and Freq. Control 35(4): 511-521

10. Okawai H, Tanaka M, Dunn F(1990)Non-contact acoustic method for the simultaneous measurement of thickness and acoustic properties of biological tissues. Ultrasonics 28: 401-410

11. Inoue H, Carnes KI, Dunn F(1993)Absorption of ultrasound by normal and pathological human gonadal tissues *invitro*. Jpn. J. Med. Ultrasonics 20(6): 349-355

12. Taniguchi N, Itoh K, Mori H(1993)Studies on ultrasonic tissue characterization - Measurement of frequency dependent attenuation and echo level of fatty liver in rats -(in Japanese). Jpn. J. Med. Ultrasonics 20(10): 574-588

13. Drewniak JL, Carnes KI, Dunn F(1989)*InVitro* ultrasonic heating of fetal bone. J. Acous. Soc. Am. 86(4): 1254-1258

14. Goss SA, Cobb JW, Frizzell LA(1977)Effect of beam width and thermocouple size on the measurement of ultrasound absorption using the thermoelectric techniqe. 1977 Ultrasonic Symposium Proceedings: 206-211

15. Parker KJ(1985)Effects of heat conduction and sample size on ultrasonic absorption measurements. J. Acoust. Soc. Am. 77(2): 719-725

16. Nyborg WL(1981)Heat generation by ultrasound in a relaxing medium. J. Acoust. Soc. Am. 70(2): 310-312

17. Yoshida Y, Inoue H, Ishida H(1994)Ultrasound velocity measurement of fatty liver model by hog tissue. The 7th Congress of World Federation for Ultrasound in Medical and Biology(Sapporo): 1-5-1

Discussion

DUNN: Thank you, Dr. Inoue. Now I ask for discussion. Dr. Ophir.

OPHIR: I have a question regarding the fat distribution. From your slide it looked like there were fat globules that were on the order of 50 or 100 microns in dimension, I couldn't really tell. But then you also said something about 10 microns. The fat in human liver perhaps is different than in pig's liver in that the fat is in micron size droplets inside the hepatocytes. Can you comment on accuracy, how your model was really constructed from that same point of view, and whether you feel that this is an appropriate model for human liver as well?

INOUE: Unfortunately, I have no data for comparing the human liver with this model. I must do such kind of analysis of the materials.

HILL: I'm not sure whether you told us the absolute temperature at which you made your measurements. I think this can be quite important, and it may explain the saturation that you saw, because the absorption coefficient, and certainly the speed of sound, in fat depends strongly on temperature. a phase change occurs at about 32^oC.

INOUE: I'm sorry, I didn't give you the temperature. The absorption measurements were done at room temperature, and the velocity measurement presented here were performed at 37^oC.

DUNN: I believe the velocity data you showed in Sapporo did provide evidence of a phase change.

INOUE: Here is the sound velocity variation at an elevated temperature. There's a substantial dependence. For absorption measurements, when the temperature is increased, the fat melts and it's difficult to measure in water. So the absorption measurements were done at room temperature. The next stage must be the controlled temperature of the materials during in the absorption measurements.

DUNN: Any further questions? Dr. Miller.

MILLER: I guess I'd first like to offer my congratulations on some very interesting work, and the velocity work is really outstanding and of genuine interest. The absorption work is coming along nicely, and I'm certain that you will sort out the questions. I'd like to add to the comment that Professor Ophir made, and that is the size of the globules. Clearly, the system that you have wisely chosen to study is inherently inhomogeneous, and Jonathan Ophir's question asks, what are the sizes of the inhomogeneities. And I believe that that is a fundamentally important question that your work and Professor Dunn's work over the years wisely addressed with the use of a phase- insensitive measurement device. By looking at temperature rise in the tissue, you are in no way influenced by the distortions of the wave fronts of the ultrasonic fields. You can appreciate this if you used a more conventional, say, piezoelectric receiving transducer. But then Dr. Ophir's question would have even another dimension, which is that smaller particles distort the field in a fundamentally different way than medium-sized or larger particles. So even apart from the biologic issue of whether human liver has its fat distributed on a certain size scale, why these important physics questions about what are the effects of distorting the ultrasonic field? However, you are independent of that with these wonderful phase-insensitive measurements. In our lab many, many years ago, we approached the same problem with an alternative approach by using a phase-insensitive acousto- electric receiver, also to avoid distortions arising from what might be called phase-cancellation effects at a piezoelectric receiver. It's an alternative technology. Each of the technologies have advantages and disadvantages. What I think is very pertinent is that we will hear two more discussions today from my colleagues Dr. Waag and Dr. Ohtsuki about the effects of the field and the inhomogeneities that the role plays. So I think it's a nice example where very practical issues are related to very fundamental studies, and I want to congratulate you and encourage you to continue this absorption work.

INOUE: Thank you very much. The size of the material in the inhomogeneity is an important part of this work. Thank you very much.

DUNN: Any further questions? If not, we now have a coffee break.

Chapter 3

Ultrasonic Characterization of Tissue from Measurements of Scattering as a Function of Angle and Frequency

This chapter reviews ultrasonic scattering research at the University of Rochester. First, models for the measurement of intrinsic scattering properties of tissue are outlined. Next, measurements that illustrate the application of these models are summarized. Then, new apparatus and planned investigations are noted.

3.1 Measurement of Scattering Properties

Important effects in scattering measurements are diagrammed in Fig. 3.1. These effects include the emitter beam pattern, the detector beam pattern, the size of the scattering volume that may be determined by the emitter and detector beams and time gates, and absorption along the incident and scattered wave paths.

A basic quantity used to describe scattering is the average differential scattering cross section. This quantity may be considered to be a function of the scattering vector K_o, which is defined as the difference between the incident wave vector I_o and the vector O_o from the scattering volume to the receiver. A representative scattering geometry is shown in Fig. 3.2.

---

*R.C. Waag, Ph.D.

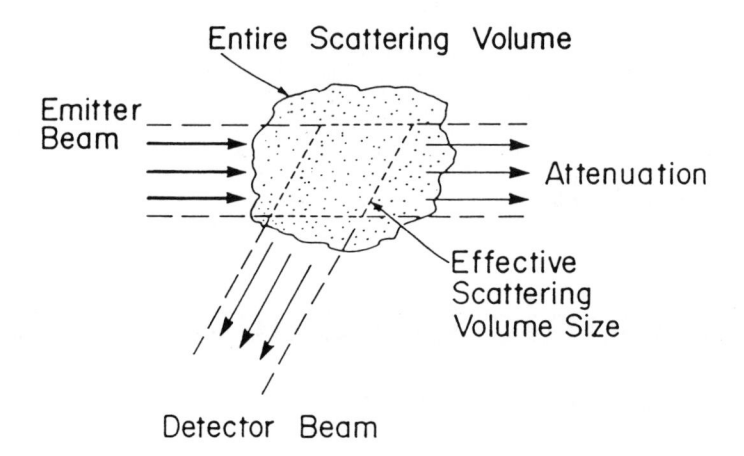

Fig. 3.1 Effects in scattering measurements

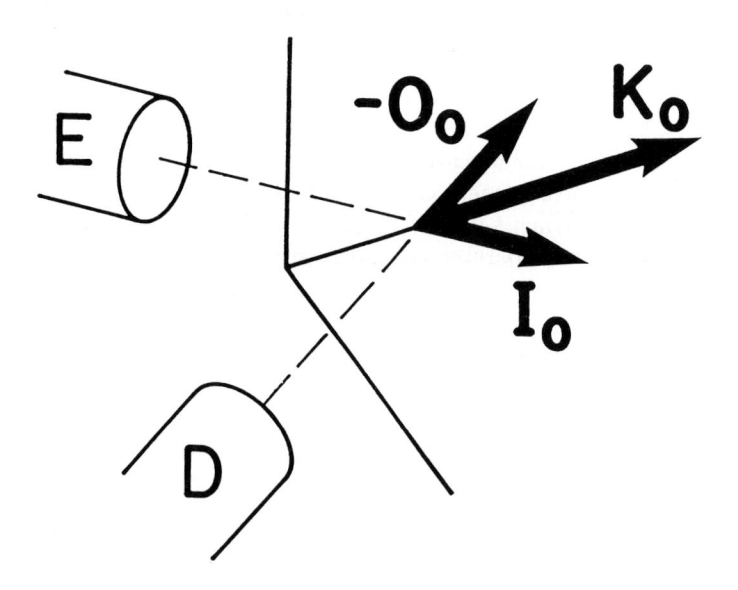

Fig. 3.2 A representative scattering geometry

3. Ultrasonic TC from Measurements of Scattering

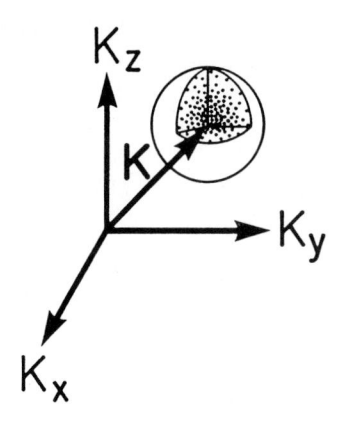

Fig. 3.3   Wavespace weighting

Textbooks give the relation between scattered pressure observed at a large distance from the scattering volume by a point detector when a plane incident wave of single temporal frequency is incident. The scattered pressure at the receiver is basically one point in a three-dimensional Fourier transform of the variations in the scattering medium. In practice, however, the relationship is more complex. Neither a single plane incident wave nor a point detector is utilized. Observations are not made in the far field of the scattering region and the signal bandwidth is not zero. Thus, a blur in the Fourier space measurement of the medium variations is introduced.

A representation of the measurement blur or weighting is shown in Fig. 3.3. A nonplanar incident wave and the finite dimension of the detector as well as a nonzero bandwidth cause the scattering vector K in this figure to have values in the region of K_o. The medium variations for all the values of K are weighted and summed in Fourier transform space by the process of measurement.

The basic relation between the measured pressure, the medium variations, and the measurement system characteristics can be written as an integration that involves a system function times the Fourier transform of the medium variations. An expression of this relation is

$$P(K_o) = \int_K \Gamma(-K)\Lambda(K, K_o)\, d^3K \qquad (3.1)$$

where P is measured pressure, Γ is the Fourier transform of the medium variations, and Λ is the system weighting function. In this expression, the system function incorporates the effects of a nonplanar incident wave field, nonzero detector size, and finite bandwidth signal.

The position of the system function in wavespace is established by the angle between the emitter and detector and also by the frequency employed. These quantities along with the emitter beam pattern and detector sensitivity pattern further determine the distribution or shape of the system weight in wavespace.

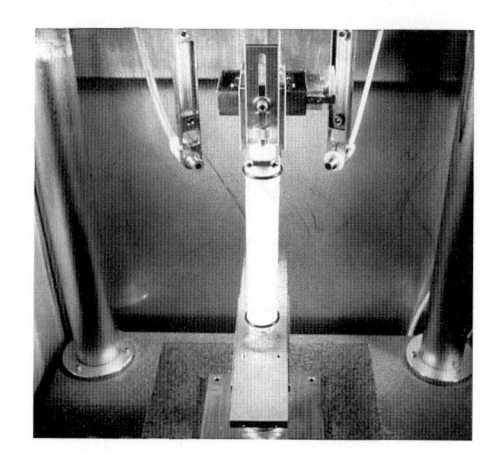

Fig. 3.4 Scattering measurement configuration

Scattering experiments should be designed in such a way that this weighting does not result in an unacceptable blurring of the the spatial-frequency properties of the tissue that is to be studied.

A convenient path for wavespace measurements varies the magnitude of the scattering vector along a fixed direction. This may be accomplished in the geometry of Fig. 3.2 by rotating the emitter and detector around the vertical axis in equal and opposite increments. The result is that the scattering vector K_o, which defines the wavespace location of the scattering vector that is the central point of the measurement, moves along the diagonal axis. The maximum distance of the scattering vector from the origin in wavespace occurs when the emitter and detector are collocated, while the minimum distance occurs when the emitter and detector are facing each other.

In summary, the measurement system weight depends strongly on geometry and apertures but a model is available for the design of scattering experiments to measure spatial frequencies of scattering medium variations in given ranges with limited blur.

3.2 Application of the Model

Measurements have been made on media for which calculations of the average differential scattering cross section can be performed. The cylindrical sample volume and two transducers – the emitter and the detector – are shown in Fig. 3.4. The transducers were rotated around their common central axis in equal and opposite directions. Since the scattering in this case is from a random medium, power was averaged to obtain a deterministic quantity. The averaging used measurements of scattered power at different positions along the axis of the cylinder.

3. Ultrasonic TC from Measurements of Scattering   35

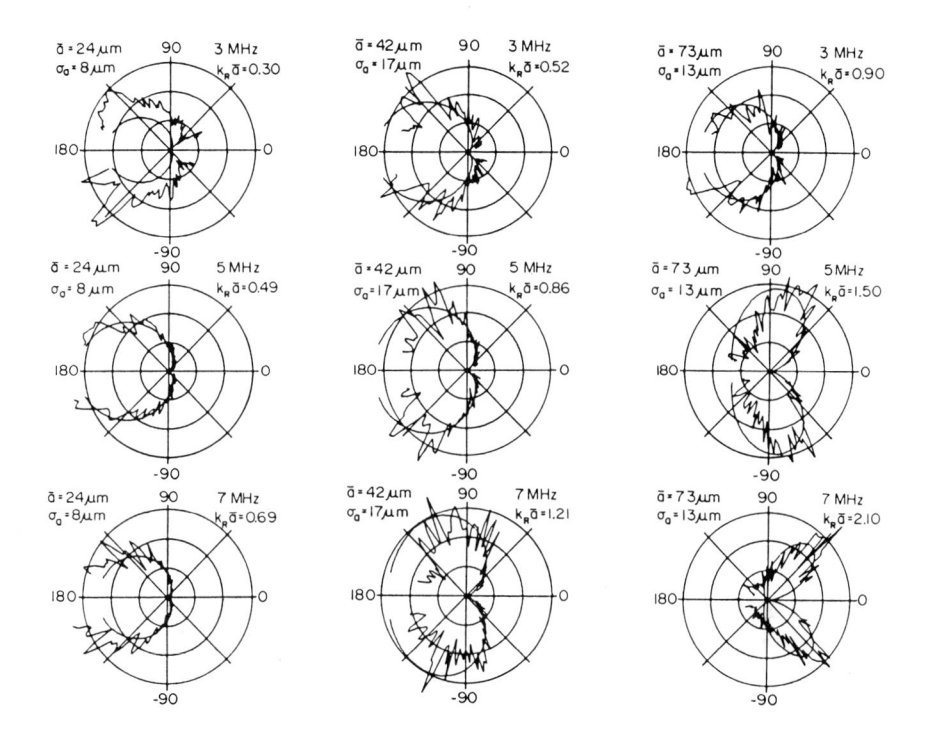

Fig. 3.5 Differential scattering cross sections for three model random media each with a different size distribution of scatters

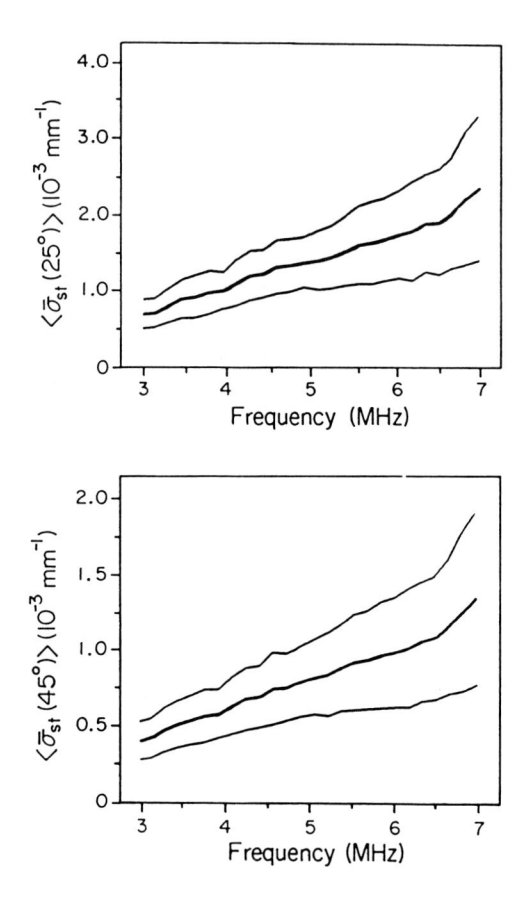

Fig. 3.8 Total differential scattering cross sections determined over two ranges of angle for calf liver

3. Ultrasonic TC from Measurements of Scattering

about proportional to the frequency to the 1.4 power. Such a dependency is quite different from the frequency-to-the-fourth characteristic that is often employed in models. Worthy of special note here is that independent measurements of attenuation using an insertion loss technique with a radiation force balance show scattering makes only a marginal contribution (2%) to total attenuation.

3.4 Contributions of Compressibility and Density Variations to Scattering

A model that describes medium variations by changes in compressibility and density can be employed to decompose average differential scattering cross section into three components. The components are the power spectrum of changes in compressibility, the power spectrum of changes in density, and the cross-power spectrum of compressibility and density changes, that is, scattering due to cross-correlation between the compressibility and density variations.

The relationships are summarized below.

$$\begin{aligned} \sigma_{sd}(\nu) &= \tfrac{k^4}{16\pi^2} S_\gamma(\nu), \nu = 2\tfrac{f}{c}\sin(\theta/2) \\ S_\gamma(\nu) &= S_\kappa(\nu) + 2S_{\kappa\rho}(\nu)\cos\theta + S_\rho(\nu)\cos^2\theta \\ S_{ij}(\nu) &= F\{B_{ij}(r)\}, B_{ij}(r) = <\gamma_j(r')\gamma_j(r'+r)> \end{aligned} \qquad (3.2)$$

In these relations, σ_{sd} is the average differential scattering cross section, S_γ is the power spectrum of scattering, ν is the spatial frequency in cycles per mm and θ is the scattering angle, S_κ is the power spectrum compressibility variations, $S_{\kappa\rho}$ is the cross-power spectrum, and B_{ij} is the correlation function of medium variation γ_i and medium variation γ_j.

The average differential scattering cross section in the above expressions is proportional to a weighted sum of the three power spectra associated with individual variations. Thus, three or more independent determinations of the average differential scattering cross section at identical values of spatial frequency but different values of scattering angle can be used to solve for values of the individual power spectra of medium variations. Assuming statistical isotropy so the measurements depend on the magnitude of spatial frequency and not on direction, steps in the process are:

1 Measure $S_\gamma(K) = S_\gamma(f,\theta)$ for a range of frequencies and angles.

2 Select data at those values of K for which there is a sufficient quantity of measurements have been made.

3 Solve an overdetermined set of equations in the presence of additive noise and statistical fluctuations.

A diagram of this process is shown in Fig. 3.9. Measurements are limited in the vicinity of the zero degrees by the main beam. A limitation also exists

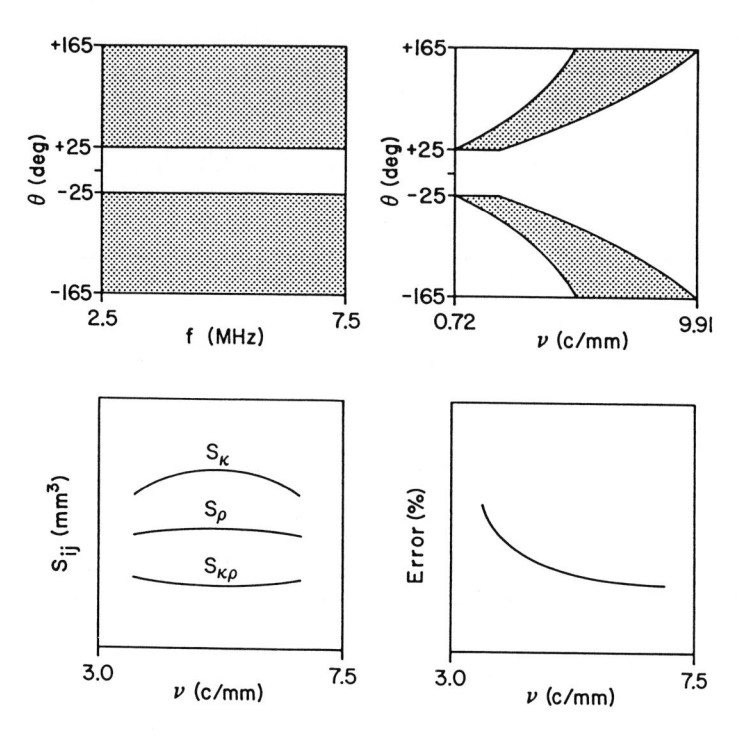

Fig. 3.9 Representation of the spectral power decomposition process

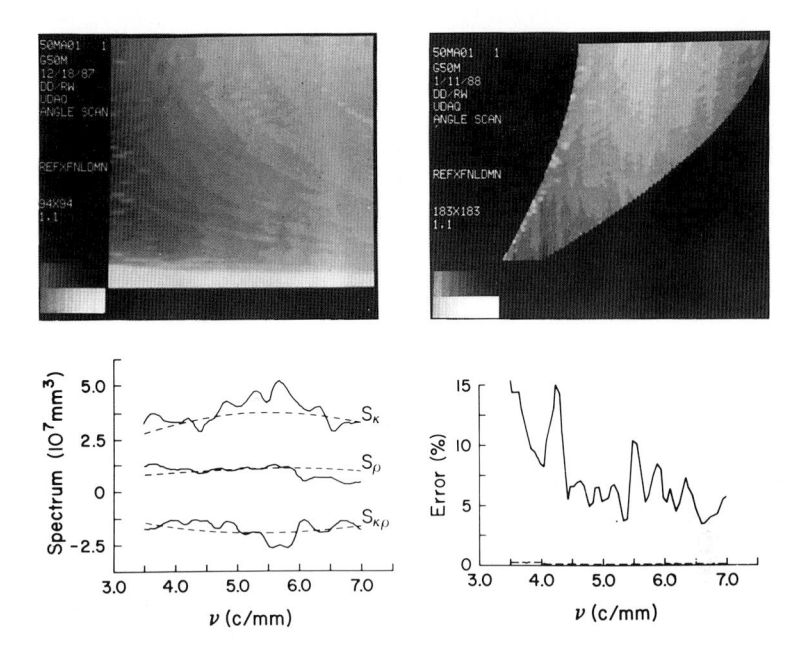

3. Ultrasonic TC from Measurements of Scattering

Fig. 3.10 Spectral power decomposition for one of the model random media (M) employed for Fig. 3.5 and Fig. 3.6

in the backscattered direction because the emitter and the detector can not be in the same portion in the apparatus described here. Measurements were made over the range of frequencies from 2.5 to 7.5 MHz. The measurements are mapped by the transformation from frequency in time to frequency in space. In the decomposition process, measurements at all available angles were used for a given spatial frequency. The equations were then solved in order to obtain the components of the power spectrum. The spatial-frequency range of the computations was determined by the available window of the measurements and by the use of a relative error criteria in a singular value decomposition.

Measurements and decomposition for the model scattering media considered earlier for which the components of the power spectra can be calculated are presented in Fig. 3.10. The decomposition from measured data are the solid lines while the decomposition calculated from theory are the dashed lines. In Fig. 3.11, analogous data based on the measurements for calf liver are shown. Significantly, in these measurements, an appreciable contribution to scattering is associated with density variations, and the cross-power spectrum is nonzero.

These results show that the power spectra of medium variations can be determined from appropriately designed scattering measurements and that density variations make an important contribution to scattering by calf liver.

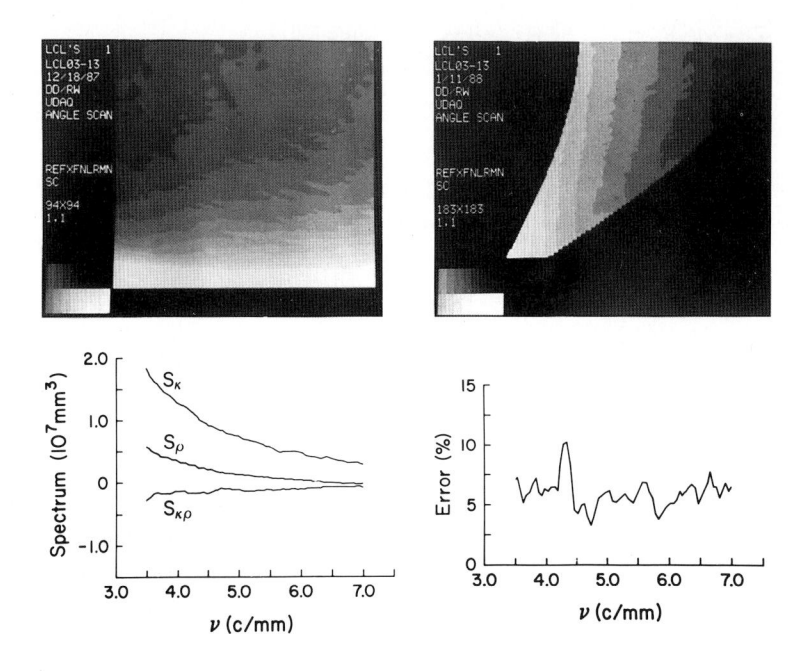

Fig. 3.11 Spectral power decomposition for calf liver

3.5 Correlation of Scattering with Tissue Morphology

Comparison of volume scattering measurements and structure observable in planar form optically through a microscope is desirable to correlate scattering properties with tissue morphology. A theoretical foundation for making such correlation has been developed at the University of Rochester. This foundation is based on the assumption that the medium is isotropic, that is, the dependence on the spatial frequencies is above described of direction. The theory relates the power spectrum of medium variations in three-dimensional space to the power spectrum that would be observed in two dimensions and also in one dimension.

Processing steps that determine from cross sectionals a power spectrum analogous to that measured acoustically are listed below.

1. Choose a cross-sectional field, L, and sampling interval, δL, to span the spatial-frequency range of interest. ($1/L \leq \nu \leq 1/2\delta L$)

2. Estimate the power spectrum in two-dimensional space by averaging squared spectral magnitudes at corresponding spatial frequencies from a number of $L \times L$ cross sections.

3. Convert the power spectrum in two-dimensional space into the power spectrum in three-dimensional space by using an intermediate power spectrum in one-dimensional space.

The results of a model study are shown in Fig. 3.12. In this study, spheres of various sizes were placed in a background medium. Representative cross sections of the spheres are given in the upper panels. A section from the distribution of larger spheres is shown in the left panel. This larger distribution can be expected to have power at lower spatial frequencies than the distribution of smaller spheres. Also shown in Fig. 3.12 are the power spectra in one, two, and three dimensions. For each distribution of spheres, the curve in three dimensions is sharply peaked while the original curve in two dimensions is less sharply peaked and the curve in one dimension is monotonically decreasing. The peaks in the distribution of spectral power associated with the larger spheres is shown to occur at a lower spatial frequency as expected.

An analogous study of pig liver is illustrated in Fig. 3.13. The upper left panel shows a representative cross section in which the collagenous septa around the liver lobules have been emphasized while the upper right panel shows the power spectrum in two dimensions. The corresponding radial distribution of spectral power in two dimensions is shown in the middle left panel and the smoothed version of that distribution is presented in the middle right panel. The one-dimensional spectrum is shown in the lower left panel while the radial distribution of spectral power that corresponds to the acoustic measurements in wavespace is shown in the lower right panel. The ring of spectral power in two dimensions and the peak in the radial distribution of spectral power in three dimensions are the result of regularity in the lobular structure of pig liver.

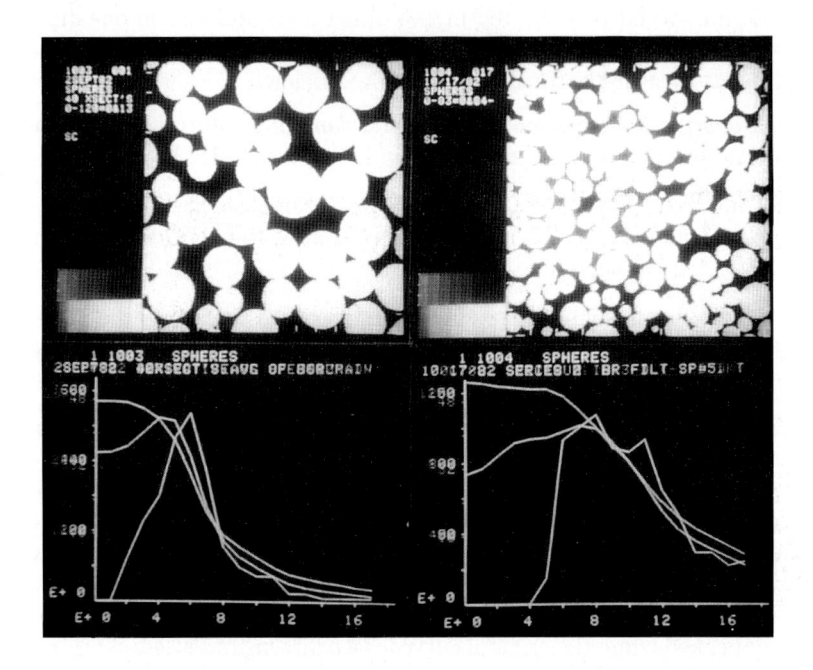

Fig. 3.12 Representative cross sections and spectral power calculations for two size distributions of spheres

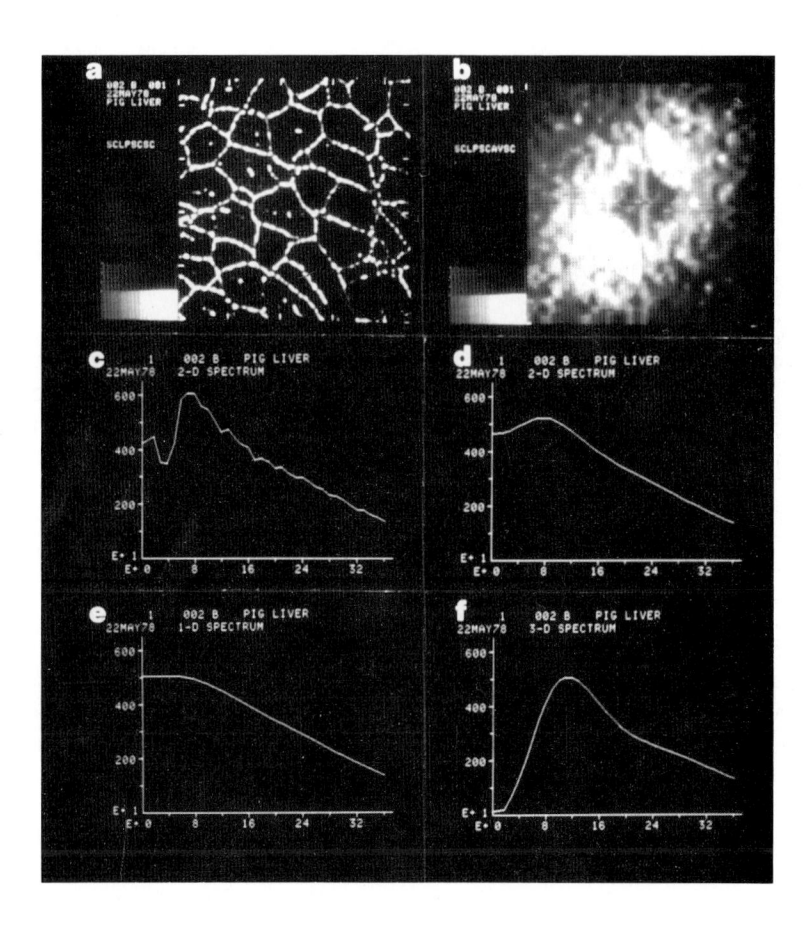

3. Ultrasonic TC from Measurements of Scattering   45

Fig. 3.13 Representative cross section and spectral power calculation for pig liver

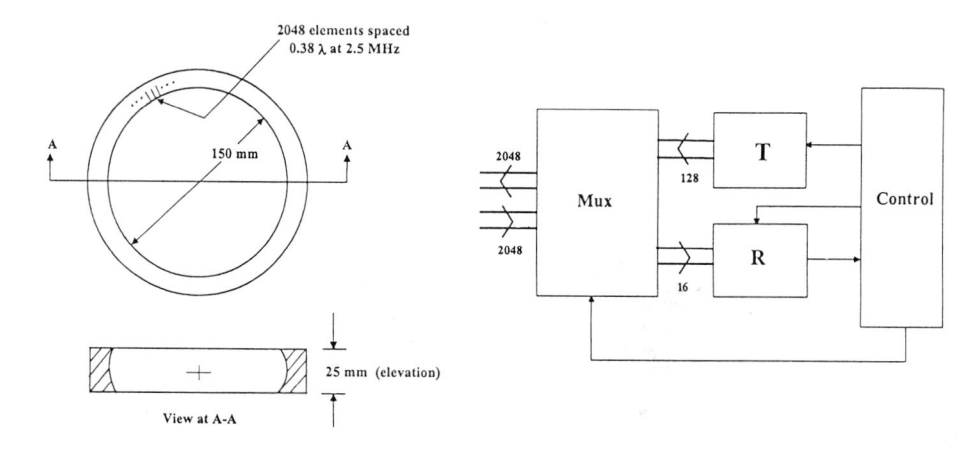

Fig. 3.14 Ring transducer and associated electronics

In summary, the cross-sectional structure of an isotropic homogeneous medium may be converted into a power spectrum that is analogous to the power spectrum measured in an ultrasonic scattering experiment and such power spectra may be used for correlation of scattering measurements with tissue structure.

3.6 New Apparatus and Additional Investigations

New apparatus is being assembled and plans exist for additional investigations. The new apparatus is comprised of a ring transducer and associated electronics. The investigations consist of scattering and imaging experiments. The new apparatus is designed to permit flexibility in the formation of emitted wave field and also in the synthesis of detector sensitivity patterns. In addition, the apparatus is configured to permit measurements of angular scattering without the need to scan transducers mechanically.

A diagram of the ring transducer and associated electronics is shown in Fig. 3.14. The diameter of the transducer ring is 150 mm. The elevation is 25 mm while the center frequency is 2.5 MHz. This relatively low center frequency was chosen because earlier measurements indicate a large amount of scattering by lower spatial-frequency structure. The pitch of the elements is 0.23 mm or 0.37λ. This results in a total of 2,048 elements around the ring, an order of magnitude more than in any other system that is typically found in practice today. A multiplexer (Mux) provides access to any contiguous 128 elements in the 2048 element ring for transmission and also provides access to any contiguous 16 elements for simultaneous reception. The transmit electronics (T) have independently programmable waveforms on each of the 128 channels. The receive electronics (R) have independently programmable time varied gain functions on each channel

3. Ultrasonic TC from Measurements of Scattering

Fig. 3.15   Who can tell the difference?

and each channel includes a 25 MHz, 12-bit A/D converter. The control electronics (Control) provide convenient access to the multiplexer, transmit electronics, and receive electronics through a Pentium-based personal computer with a large amount (10 gigabytes) of disk storage. The electronics permit synthesis of an aperture of up to 2048 elements in sets of up to 128 elements for transmission and a receive aperture of up to 2048 elements in groups of up to 16 elements.

For tissue characterization, high spatial-frequency resolution measurements of scattering as a function of angle and frequency are planned for normal and diseased tissue. The measured scattering characteristics will be compared to predictions based on tissue morphology. Also, these measurements of scattering will be decomposed into a power spectra due to compressibility and density variations.

3.7   Conclusion

The ability of researchers to characterize tissue ultrasonically is not as good as the dolphin that can acoustically tell the difference between a savory McDonald's hamburger and a worm on a dangerous hook as depicted in Fig. 3.15 but progress has been made. Models are available to guide ultrasonic measurements of the intrinsic scattering characteristics of tissue and correlation with tissue morphology. Measurements of scattering from tissue-like media and also from calf liver have illustrated that intrinsic scattering characteristics can be measured by appropriately designed experiments. New apparatus that is becoming available and additional investigations are planned with the apparatus to develop a scattering characterization of tissue.

Discussion

DUNN: Dr. Waag's presentation is open for discussion. Dr. Hill.

HILL: Well, I'd really like to congratulate Bob Waag on a really outstandingly brilliant piece of work. I really want to make just two points. We were discussing in the last session when I was asked what the future ideal equipment would be. I think Bob Waag has really shown us what I would choose as being the way to go in order to get the best chance of characterizing tissue on the basis of fundamental structure, which is really what it's all about. I know very well, because we've been interested in this field ourselves, that this is a very difficult field in which to work. It seems to me that it has tremendous promise for the future.

The other comment I wanted to make, and it's something that I've thought about over the years and it's one of these things that you've never got round to doing, but it seems to me that there are very interesting parallels between the kind of thing you've been talking about and the very long-established field of X-ray crystallography, particularly where it's applied to powder crystallography and crystallography of structures that are not totally regular. It seems to me that it would be a very nice thing at some stage in the future to try to bring together these two fields and to at least review the extent to which they complement each other and can assist each other and the ideas that they can take. I don't know whether you've had any thoughts about that yourself. But anyhow, great congratulations.

WAAG: Well, thank you very much for your kind words. I can say that we've mortgaged our soul. I don't know how that translates into Japanese, but we have mortgaged our soul for this transducer and electronics, and we are, though, nevertheless excited by the prospect of doing these measurements in a very efficient and careful way. We have thought a little bit about the relation between the acoustic determination of tissue structure based on scattering, and the relation of that to the X-ray or other particle-scattering methods for characterizing materials. And I think we're inspired by the fact that there's a history of Nobel prizes in these other fields for the application of the methods in the characterization of materials. We're impressed that there has been relatively little study using these wave-based methods in acoustics. I know that Professor Hill's group started doing these measurements years ago about the same time that we did. There haven't been many individuals look at the angle and frequency dependence. Most of the applications have been using backscatter, which is definitely limited. It's not possible through a backscatter measurement to separate the contributions associated with compressibility and density. I just don't know a lot about the X-ray techniques, but I'm impressed that one of the things we would like to get in our measurements is something akin to the pair correlation function that the physicists use to characterize the random nature of materials, non-biological materials. Also, I'm impressed that the basic principles are the same for us, but the conditions of the measurements are quite different. The reason I think that one doesn't find in the textbooks some of these ultrasonic relations I described is that, first of all, in X-ray the plane-incident wave and the far-field assumptions are very good. There not very good in acoustics. And so the experimental design must be done a lot more carefully. In X-ray, it's possible to go to a store and buy an X-ray diffractometer. In our case, we can't do that. We have to build our own, and of course this is a challenge and a problem all by itself. That's maybe a brief response to your comment, Professor Hill.

MILLER: Let me begin by adding my congratulations to those of Kit Hill. I remember at the earliest tissue characterization meetings in Washington approximately 20 years ago the early work that Professor Waag presented in exactly this way, and I think there's a lesson for us all, which is that it may take 20 years of very hard work to yield the kinds of accomplishments we see here today. So my congratulations are most sincere, and they have a great deal to do with saying that sticking to the topic is something that all of us need to remember. I will say that by long tradition our Japanese friends have done much better than my U.S. colleagues. In the United States, we have a dreadful history of losing interest in things much too soon, and Bob Waag is to be congratulated for staying with it. I think the Japanese as a group of people have a very long willingness to stay with things and really see them through to conclusion.

I have a technical comment, and one that I hope might be of interest. Bob Waag has very nicely shown us how to decompose the angle-dependent scattering into measurements of changes in density and changes in compressibility. As Professor Waag said, compressibility variations in homogeneity in the tissue resulting from variations in compressibility give rise to a monopolar-like structure in the scattering. Density variations in the tissue give rise to a dipolar-like component, and it's the combination of those two that gives rise to the result, and I'm simply restating what Dr. Waag said. There is, however, another decomposition that our lab published a few years ago that isn't so widely known but might be of interest to the audience, and that is in addition to decomposing according to changes in density and changes in compressibility, if the contrast is low, and that's exactly the case throughout soft-tissue ultrasound, so liver, for example, is a relatively low-contrast medium. If the contrast is low, then you can approximate the very precise expressions that Dr. Waag has with some almost as precise, very good approximations. And they roughly go as follows: changes in acoustic impedance give rise to a backward-directed cardioid shape, and changes in acoustic velocity give rise to a forward- directed cardioid shape. Now, you might say why go to a more complicated shape as opposed to a monopole or dipole. Here's the reason. Suddenly, there's a very simple interpretation of Bob Waag's forward-directed dominance in the liver case. When the forward-directed signal is large, it means that local variations in velocity are dominating. When the reverse is true, namely, that when backscatter-like signals are large, then it's local variations in impedance which are dominating. Now this is entirely equivalent to the breakdown based on density and compressibility. It's just an alternate way. But the nice feature is that it's exactly true at 180 degrees, which is backscatter. At 180 degrees in backscatter it is only changes in impedance that matter in terms of producing the backscatter. It is also exactly true in zero degree forward scatter, where only changes in velocity control the outcome. So this decomposition is an approximation, but a very good one for soft tissue, and I believe it allows us to look at your liver data and see that local variations, local inhomogeneities, local lumpiness arising more from velocity than impedance give rise to the forward-directed scattering.

DUNN: Any other further remarks? Dr. Jones.

JONES: I just wanted to follow up on one comment on what Dr. Miller said, and that is if you look at 90 degree scattering, then the 90 degree scattering is only due to fluctuations in compressibility. So backscatter you can refer to as due to fluctuations in impedance, and forward scattering is due to fluctuations in velocity. But if you look at 90 degrees, you can separate out the fluctuations in density compressibility and only look at fluctuations and compressibility at 90 degrees.

WAAG: Well, I'd like to just thank the participants, Dr. Miller, and Dr. Jones for those interesting comments. I think they are a propose, and they do provide useful information. I could simply add briefly that the relation between the impedance variations and the velocity variations and the compressibility and density variations is easily derived for those who aren't so familiar with these relations by noting that impedance is ρc, the density times the sound speed. And then we can substitute for the sound speed, for instance, its expression in terms of the density and compressibility. We can take the differential, then, of that quantity and show the relations that were described here today. So those are available in Professor Miller's paper. Actually, we wrote a paper some years ago about that, too, but we didn't push it. So thank you very much.

DUNN: Further comments? Dr. Tanaka.

TANAKA: Thank you very much for showing a very interesting and informative results of your survey. We ourselves are engaged in echocardiography, and the problem of backscattering is something which affects the imaging. And we have put a lot of focus and attention upon your investigation for many years. After listening to your talk, I think a lot of information has been given to us. And your presentation elucidated a lot of questions we have been holding all the time.

There's one point on which I'd like to ask for your comment. That is, in clinical areas, after computing acoustic impedance and after that the intensity of reflection is predicted. And in that case, there is a problem of the shape and size of the reflecting body. How these are considered to be the parameters that should be taken into the calculation–that is a very practical problem. In Japan, we often talk about these models, and we usually use spherical models in calculation. But when it comes to the actual boundaries of the tissues, there are various kinds of tissues. Therefore, the borders are not regular, but irregular. Therefore, the shape of the reflector is very complex and varies in size and dimensions. And the power of reflection might be different depending upon those situations. And what kind of parameters to be used and how these parameters can be taken into the calculation–do you have any comments on this point? And the other point is the strength of backscattering. When we try to make a judgment as to the intensity of backscattering, of course attenuation of the pathway should be considered. But in addition to that, the reflectivity of the reflecting body itself should be taken into consideration. But to what extent should this be taken into consideration? Could you comment on this point.

WAAG: Thank you very much, Professor Tanaka, for your comments. I'm embarrassed to say that we started working in the area of cardiac tissue characterization many years ago, but soon we learned how difficult this field is, and we moved away. You'll hear, I think, some very interesting and impressive work from Dr. Miller later today about quantitative characterization of cardiac tissue. This is not an area where we have done any recent research. But I can comment briefly about your question or statements about the reflectors. The work I described basically presumed that the scattering volume is limited by the beams and the time gates, and not by the surface of the cardiac wall, the endocardium or the epicardium. Our assumption that the beams limit the volume instead of the sample limiting the scattering volume makes one problem go away–the problem of reflection from the boundaries. So it's another complexity to put in those boundaries. And we have not done that yet.

DUNN: Any further discussion? If not, we thank you very much, Dr. Waag.

Chapter 4

Sound Field of Disk and Concave Circular Transducers

4.1 Introduction

Living tissue is inhomogeneous media for ultrasound propagation. For example, sound speed in tissue depends on the structure and the contents of materials such as collagen and fat. Thus, numerical values of the acoustical parameters can be helpful for medical diagnosis.

When we are going to measure the acoustic parameters of a small volume of tissue, received signal has two kinds of information. One is on the path from transducer to the small volume around the observed point. The other is on the small volume around the observed point in tissue.

In case of the measurement of excised tissue in water, ultrasound propagation is in homogeneous media, that is, water. So focused area in tissue can be estimated by the sound field in homogeneous media.

Here the theoretical estimation method of ultrasound field of a disk or a concave circular transducer is introduced. This is named as Ring function method.[2]

4.2 Expression of Sound Pressure Emitted from a Transmitter with Ring Function

A plane transmitter with infinite buffle vibrating at uniform velocity $v(t)$ emits sound into non absorptive homogeneous media and forms a sound field. Lord Rayleigh derived a following expression for pressure p at an observing point O

---
*S. Ohtsuki, Ph.D.

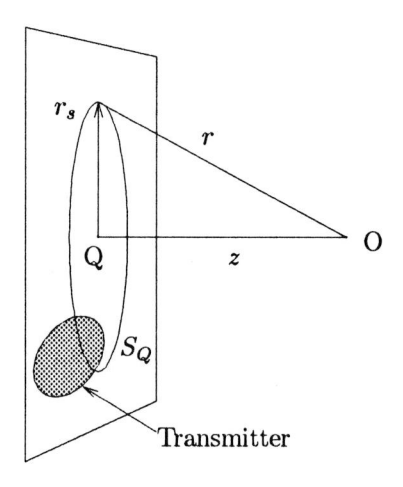

Fig. 4.1 Plane transmitter and observing point

in the field:[1]

$$p = \rho \int_{S_T} v'\left(t - \frac{r}{c}\right) \frac{dS}{2\pi r} \quad (4.1)$$

where ρ and c are the density and the sound speed of the media respectively, r is the distance from an area element dS on the surface S_T of the transmitter to the observing point O, and prime for v denotes time derivative. Above equation can also be used approximately in cases of transmitters with curved surface, if the dimension of vibrating surface is smaller than its radius of curvature.

Here vibrating area elements at a range r from an observing point are integrated considering wave propagation in homogeneous media. Then we can get a one-dimensional integral expression in stead of two-dimensional integral expression, Eq. (4.1).

In case of a transmitter with a plane or spherical surface, Eq. (4.1) can be transformed into a convenient one-dimensional integral of r by using ring function.

4.2.1 One-dimensional Integral Expression for a Sound Field Formed by a Plane Transmitter

Figure 4.1 shows the locus of points on a plane at the distance r from an observing point O. The locus is a circle of radius r_s and its center Q is the intersection point of the plane and the perpendicular to the plane through the point O. The area S_Q bounded by the circle is given as

$$S_Q = \pi r_s^2 = \pi(r^2 - z^2) \quad (4.2)$$

where z is the distance OQ.

4. Sound Field of Transducers

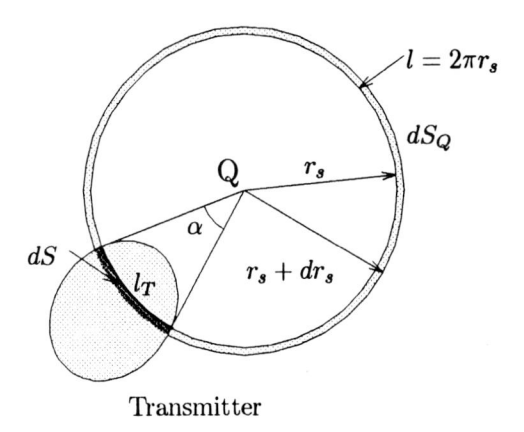

Fig. 4.2 Element of area

Regarding z as a constant and differentiating, we get

$$dS_Q = 2\pi r dr \tag{4.3}$$

The area dS_Q is illustrated in Fig. 4.2. Then, the area element dS in Fig. 4.2 is given as

$$dS = \frac{\alpha}{2\pi} dS_Q \tag{4.4}$$

where α is a function of r.

Let us define the ring function $R(r)$ as

$$R(r) \equiv \frac{\alpha}{2\pi} \tag{4.5}$$

The ring function $R(r)$ can also be expressed as the ratio l_T/l in Fig. 4.2 as

$$R(r) = \frac{l_T}{l} = \frac{l_T}{2\pi r_s} = \frac{l_T}{2\pi \sqrt{r^2 - z^2}} \tag{4.6}$$

When the circle is contained in the transmitter, $R(r)$ is unity. With this ring function, we can transform Eq. (4.1) into

$$p = \rho \int_{r_2}^{r_1} v'\left(t - \frac{r}{c}\right) R(r) dr \tag{4.7}$$

where r_1 and r_2 are the minimum and the maximum distances from an observing point O to a transmitter, respectively.

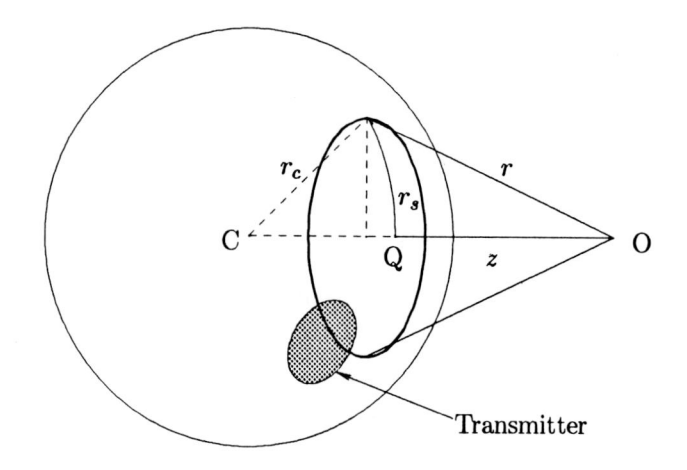

Fig. 4.3 Spherical transmitter and observing point

4.2.2 One-dimensional Integral Expression for a Sound Field Formed by a Spherical Transmitter

Figure 4.3 shows a circle on a sphere as the locus of points distance r from an observing point O. The area S_c inside the circle is given as

$$S_c = \pi r_c \frac{r^2 - z^2}{r_c + z} = \pi r_c \frac{r^2 - (r_c - z_c)^2}{z_c} \tag{4.8}$$

where z is measured from the point Q, z_c is measured from the point O, and z is shown as positive in Fig. 4.3. Differentiating Eq. (4.8) with respect to r, on the condition that r_c, z and z_c are constant, we get

$$dS_c = 2\pi r \frac{r_c}{r_c + z} dr = 2\pi r \frac{r_c}{z_c} dr \tag{4.9}$$

Now, let us define K as follows:

$$K \equiv \frac{r_c}{r_c + z} = \frac{r_c}{z_c} \tag{4.10}$$

Using Eqs. (4.5),(4.9) and (4.10), an area element dS of the transmitter can be expressed as

$$dS = R(r)dS_c = 2\pi r K R(r) dr \tag{4.11}$$

Replacing dS in Eq. (1) with the above result, we get

$$p = \rho K \int_{r_1}^{r_2} v'\left(t - \frac{r}{c}\right) R(r) dr \tag{4.12}$$

For a plane transmitter, r_c is infinite and K is unity. Then, for this case, Eq. (4.12) is identical with Eq. (4.7).

4. Sound Field of Transducers

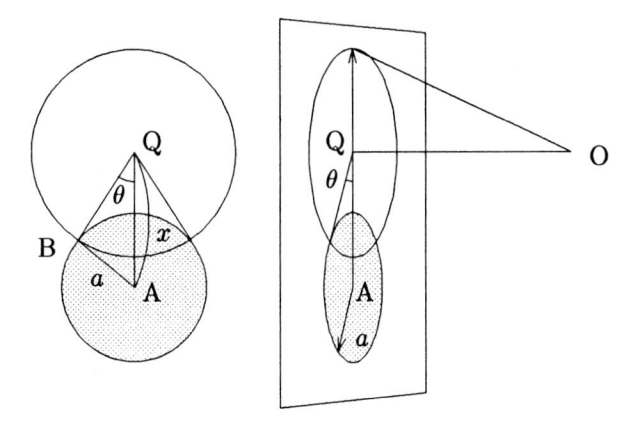

Fig. 4.4 Ring function in the case of a disk transmitter

Let $v(t)$ sinusoidal velocity as $V\sin\omega t$, Eq. (4.12) becomes following expression in complex.

$$P = jkK\rho cV \int_{r_1}^{r_2} e^{-jkr} R(r) dr \qquad (4.13)$$

where k is the wavelength constant $\left(= \dfrac{\omega}{c} = \dfrac{2\pi}{\lambda}\right)$.

In Fig. 4.3, the transmitter has a convex vibrating surface. For a concave transmitter, it is convenient to reverse the direction of z by reversing its sign.

4.3 Some Examples Using the Ring Function

Usually we use transmitters with linear or circular boundaries or a combination of them. Ring functions for such transmitter types can be given by the combination of ring functions for lines and for circles. Hence, let us find ring functions for transmitters with circular boundaries. The ring functions for transmitters with linear boundaries can be derived more easily.

4.3.1 Ring Function for a Disk Transmitter

Now, let us examine the case of a disk transmitter of radius a shown in Fig. 4.4. By definition, the ring function $R(r)$ is given as

$$R(r) = \frac{2\theta}{2\pi} = \frac{\theta}{\pi} \qquad (4.14)$$

Applying the law of cosines for the triangle to $\triangle$ ABQ, we get

$$\cos\theta = \frac{x^2 + r_s^2 - a^2}{2xr_s} \qquad (4.15)$$

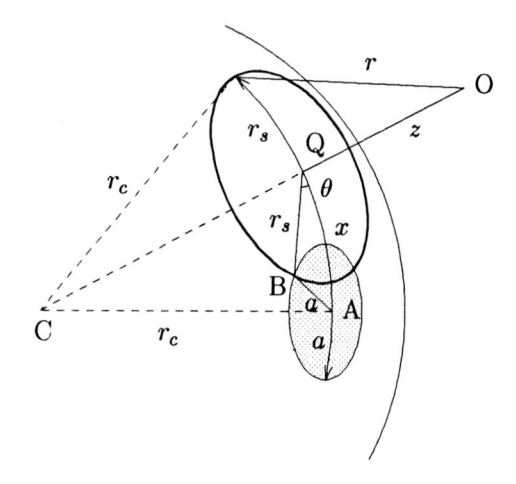

Fig. 4.5 Ring function in the case of a transmitter with spherical surface

From Eqs. (4.14) and (4.15), and Fig.4.1, we get

$$R(r) = \frac{1}{\pi} \cos^{-1} \left(\frac{x^2 - a^2 - z^2 + r^2}{2x\sqrt{r^2 - z^2}} \right) \quad (4.16)$$

4.3.2 Ring Function for a Spherical Transmitter with Circular Boundary

Let us examine the case of a convex transmitter as shown in Fig. 4.5. Applying the law of cosines of the spherical triangle △ ABQ, the ring function is expressed as

$$R(r) = \frac{1}{\pi} \cos^{-1} \left(\frac{\cos \frac{a}{r_c} - \cos \frac{r_s}{r_c} \cos \frac{x}{r_c}}{\sin \frac{r_s}{r_c} \sin \frac{x}{r_c}} \right) \quad (4.17)$$

where

$$r_s = r_c \cos^{-1} \left(\frac{r_c^2 + (r_c + z)^2 - r^2}{2r_c(r_c + z)} \right) \quad (4.18)$$

4.3.3 Sound Pressure on the Axis of the Transmitter

When the observing point is on the axis of a circular transmitter, the value of the ring function $R(r)$ is unity in the range of r from $r_1 = z$ to r_2 (from the observing point O to the edge of the transmitter), otherwise it is zero. That is:

$$R(r) = \begin{cases} 1 & z \leq r \leq r_2 \\ 0 & elsewhere \end{cases} \quad (4.19)$$

4. Sound Field of Transducers

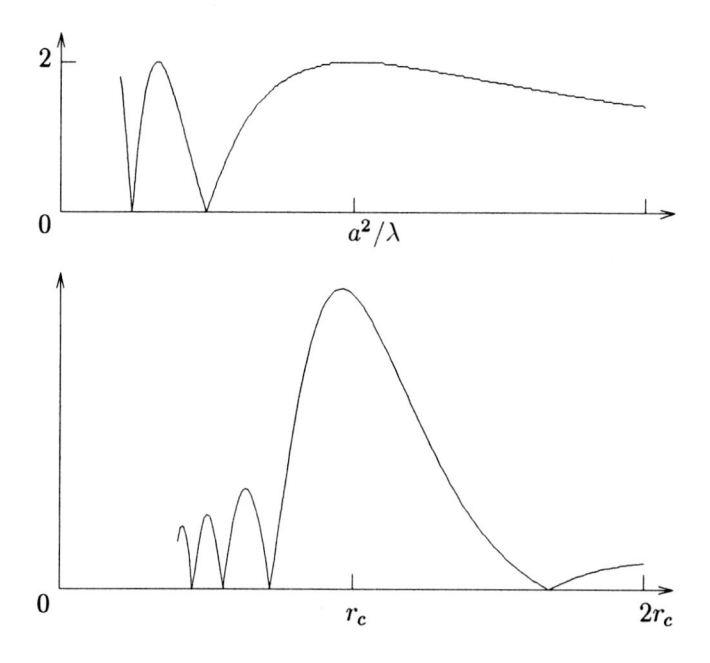

Fig. 4.6 Sound pressure on the axis of transmitter

Then eq.(4.13) with eq.(4.19) becomes

$$\begin{aligned} \boldsymbol{P} &= jkK\rho cV \int_z^{r_2} e^{-jkr} dr \\ &= K\rho cV(e^{-jkz} - e^{-jkr_2}) \\ &= K\rho cVe^{-jkz}(1 - e^{-jk\delta}) \end{aligned} \quad (4.20)$$

where $\delta \equiv r_2 - z$.

The magnitude of sound pressure on the axis of transmitter is

$$\begin{aligned} P &= |\boldsymbol{P}| \\ &= K\rho cV|1 - e^{-jk\delta}| \\ &= 2K\rho cV \left| \sin\left(\frac{k\delta}{2}\right) \right| \end{aligned} \quad (4.21)$$

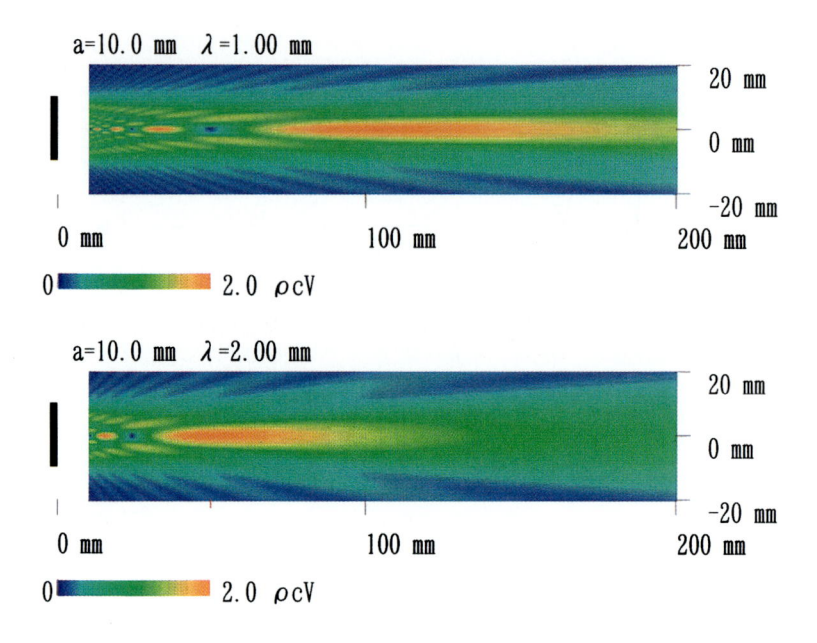

Fig. 4.7 Examples of sound field of disk transmitters

4.4 Examples of the Sound Field

4.4.1 Disk Transmitter

Two examples of sound field of disk transmitters are shown in Fig. 4.7, last maximum position along the axis is approximately a^2/λ. So distance in the axial direction can be normalized by a^2/λ and the distance normal to the axis can be normalized by the radius a of the disk. Then normalized sound field of disk transmitter can be obtained.

4.4.2 Concave Circular Transmitter

Three examples of sound field of concave circular transmitters are shown in Fig. 4.8 In case of a concave circular transducer, the distance along the axial direction can be normalized by the radius of curvature R_c and the distance normal to the axis can be normalized by the radius a of the concave circular transmitter. The sound field pattern is classified by the ratio of a^2/λ to R_c.[3]

4. Sound Field of Transducers

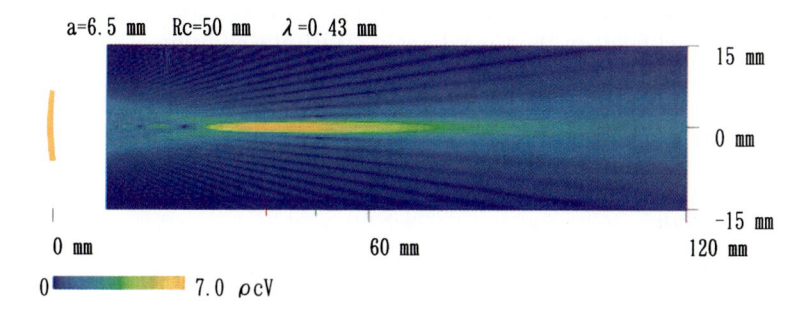

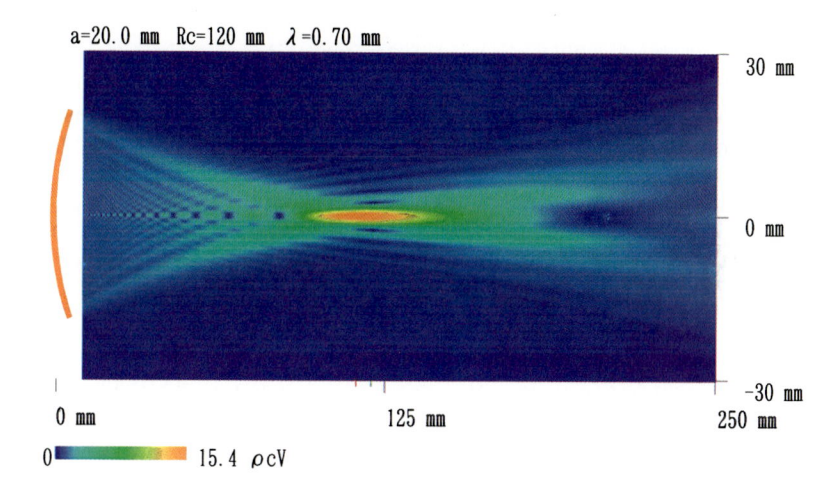

Fig. 4.8 Examples of sound field of concave circular transmitters

4.5 Conclusion

In order to calculate easily the sound field generated by a transmitter, a one-dimensional integral expression has been derived from Rayleigh's expression by using the ring function defined here. As examples of sound fields generated by a circular concave transmitter excited by a continuous sinusoidal wave, and one generated by a disk transmitter were shown.

This method of calculation using ring functions can be applied to a plane or to a spherical transmitter. The ring function method is suitable for calculation of pulsed ultrasound fields in the time domain. With the addition of some terms to account for loss of energy during propagation, this method can be applied to the case of absorptive media. Moreover, this method can be extended to describe sound fields in inhomogeneous media.[4]

References

1. Lord Rayleigh: Theory of Sound, Vol. 2 (1945), p. 107, Dover Publ.

2. Shigeo OHTSUKI:"Ring Function Method for Calculating Nearfield of Sound Source", Bulletin of the Tokyo Instituded of Technology, No.123(1974), pp.23-31.

3. Motoyoshi OKUJIMA and Shigeo OHTSUKI:"Approximate Estimation of Beam Width of Forcusing Transducer and Variable Aperture Transducer", Jarnal of Japanese Society of Ultrasound in Medicine, Vol.5, No.2(1978), pp9-14.

4. Haitao PAN and Shigeo OHTSUKI:"Shell Function Method for the Calculation of Sound Fields in Inhomogeneous Media", JASA(1994)

Chapter 5

Non-Contact Measurement of Sound Speed of Tissues

There are about 15,000 members in the Japanese Society of Ultrasound in Medicine, and the study of ultrasonics and its application in medicine are greatly pursued today. The reason for great activity is that manufacturers and their engineers are developmental work, and in the medical field the physicians are using these instruments continuously. The manufacturers and the engineers are on one side of the screen in which the instrument is made. On the other side of the instrument, the doctors are watching the screen. Sometimes their understanding of the image on the screen is different than imagined by the engineers. The medical ultrasound science is the common ground for both groups. Thus the doctors, physicians, engineers, and physicists must come together to have a common ground in developing instruments for these uses.

One field of medical ultrasound science is the detection of tissue information. There are various multimedia, like MRI and X-rays, that can be utilized for this purpose. With that as a background, we have to collect information on living tissue on ultrasound.

The relationship between the speed of ultrasound measured by the non-contact measurement system and the physical change estimated from optical microscopic figures and density will be introduced here. The non-contact measurement system of sound speed uses several megahertz pulses with fresh tissue in vitro. The system accuracy is about 0.2 %. That is about 3 m/s of sound speed in tissues.

Up to now, the sound speed of fresh and formalin fixed rat liver, heart, and kidney, and formalin fixed and frozen human liver, heart, and kidney were measured. The measurement results of the liver of rat, normal and diseased, will be shown here.

---

[*]H. Hachiya, Ph.D.,Dr.MS. and S. Ohtsuki, Ph.D.

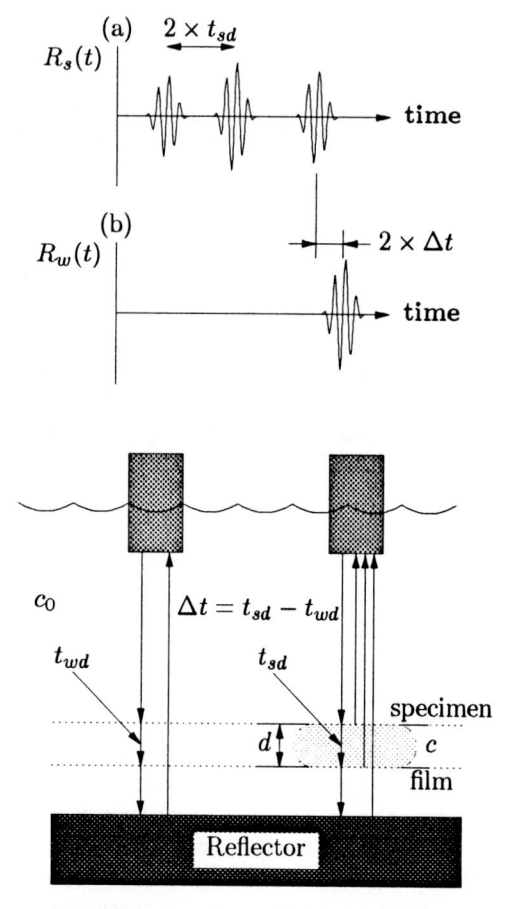

Fig. 5.1 The configuration of measurement

5.1 The Measurement System

Figure 5.1 shows the configuration of measurement[1)2)]. The tissue sample is placed on a polymer film in a liquid medium having speed of sound c_o. Another polymer film is contacted at the upper side of the tissue. t_{sd} is the travel time between the front and the rear interface of the sample; that is, the travel time of sound passing through the thickness d of the sample. Δt is the difference in the travel time from the transducer to the reflector with and without the sample placed on polymer stage. From these values, the sound speed of sample c and thickness d are given by the equations

$$c = c_o \left(1 - \frac{\Delta t}{t_{sd}}\right) \tag{5.1}$$

$$d = c_o(t_{sd} - \Delta t) \tag{5.2}$$

These values, t_{sd}, or Δt, are estimated not in the time domain, but in the frequency domain considering multipath propagation in non-uniform tissues. This method does not require physical contact of the ultrasonic probes to living or freshly excised tissue specimen. The estimates of Δt and t_{sd} utilize all the information contained in the waveform obtained from the frequency-time analysis of the received signal. Thus, the least mean error of sound speed is less than 3 m/s.

5.2 Sample Preparations

Table 5.1 shows groups of the sample preparations. Male rats of the Wister strain weighing 200–650 g were used in these measurements. The rats were divided into four groups. One is the normal group, which is the control group. The other three groups were administered carbon tetrachloride (CCl_4) to cause liver disease[4]. The normal group consists of 24 normal rats. The second group was 6 rats which were administered CCl_4 at the rate of 0.1 ml/kg of body weight twice a week for 6 to 8 weeks. CCl_4 was mixed with 9 volumes of olive oil. The mixture was administered at a dose of 1 ml/kg by intramuscular injection. This group is labeled as $CCl_4(A)$. The third group was 7 rats which were administered CCl_4 at the rate of 0.5 ml/kg only once, and the livers excised 24 to 72 h after injection. CCl_4 was mixed with an equal volume of olive oil. The mixture was administered at a dose of 1 ml/kg. The last group was 9 rats, which were administered CCl_4 at the rate of 0.5 ml/kg twice a week for the period of 3 to 17 weeks. For the groups given CCl_4, their only drinking water was tap water in which phenobarbital was dissolved at the concentration of 0.5 g/l to accelerate the progress of the disease.

The specimen to be measured was placed in the water tank which was filled with 0.9% saline at 37 °C , as a coupling medium. The ultrasonic pulse was transmitted from the transducer to the sample, and to the brass reflector behind the sample. The reflected wave was received by the same transducer. The received signal was captured by a waveform recorder at a 20 MHz sampling rate, and transferred to the computer to determine the times, t_{sd} and Δt. The transducer used had a 3.5 MHz center frequency, and 13 mm diameter and 50 mm focal length. The beam width around the focal length is 1 mm (-3dB). The speed of sound was determined at the 25 points by scanning a 4 mm by 4 mm area at

Group	CCl_4	Period	No. of Rats
Normal	None		24
$CCl_4(A)$	0.1 ml/kg ; twice a week	6-8 weeks	6
$CCl_4(B)$	0.5 ml/kg ; only once	24-72 h	7
$CCl_4(C)$	0.5 ml/kg ; twice a week	3-17 weeks	9

Table 5.1 Dosage schedule

1 mm intervals. Rat liver was placed in 37 °C saline within 5 min after excision, and left on the stage about 15 min to obtain a uniform temperature distribution at the temperature of the outer liquid medium. The complete procedure of measurement was finished within 30 minutes of the excision.

The density measurements were carried out by placing a piece of tissue sample into a graded series of copper sulfate($CuSO_4$) solutions of known specific gravities, and noting whether the piece rose or fell in the solutions[5]. The copper sulfate solutions were prepared with densities spaced 0.002 g/cm^3 apart.

5.3 Measurement Results

Figure 5.3 shows the relationship between the measured sound speed of liver and the body weight of the control normals. The closed circles and vertical bars are, respectively, the average and the standard deviation of the sound speed of 25 scanning points of each rat liver. The sound speed of rat liver 7 weeks old having a minimum weight (208 g), and 28 weeks old having maximum weight (628 g) are 1604.6±3.6 m/s and 1603.2±3.8 m/s, respectively. The standard deviation of the measured sound speed in normal rat liver is almost the same as the result of the preliminary experiment[3]. There is little difference in the sound speed depending on the measured position. The maximum sound speed and minimum sound speed of a normal specimen are 1611.2 ± 3.6 m/s and 1598.4 ± 7.6 m/s, respectively. Thus, the individual differences of sound speed in liver tissue are less than 1%. The average and standard deviation of the sound speed of 24 normal liver specimens are 1605.1 m/s and 3.2 m/s, which is 0.2% of the average sound speed. So the individual differences are very small and the measurement is found to be valid for precise measurement of tissues. The sound speed of the normal rat liver is not related to body weight or age.

Figure 5.4 shows the sound speed versus density of normal rat liver and heart tissues. The sound speed of a normal heart is less than the rat liver, but the individual difference is small – the same as the rat liver. The average sound speed is about 1580 m/s. The density variation between individuals is also small.

Figure 5.5 shows the relationship between density and speed of sound in all of the liver specimens. The closed circles, open circles, open squares and open triangles correspond to speeds of sound in normal, fatty(CCl_4(A)), acutely injured(CCl_4(B)), and cirrhotic livers(CCl_4(C)), respectively. The vertical bars are represent the standard deviation of the sound speed for 25 scanning points of each liver specimen. The standard deviation of density for each rat liver specimen was not determined, since many pieces for density measurement were not used. But it is considered that the maximum error in density was less than 0.002g/cm^3, which was the difference between the prepared $CuSO_4$ solutions.

Figure 5.2 shows the microscopic section of the HEstained liver tissue specimens of the normal and CCl_4 group rats. Histologically, the stained liver tissue specimens of CCl_4(A) were recognized to be indicative of fatty liver. In the

5. Non-Contact Measurement of Sound Speed

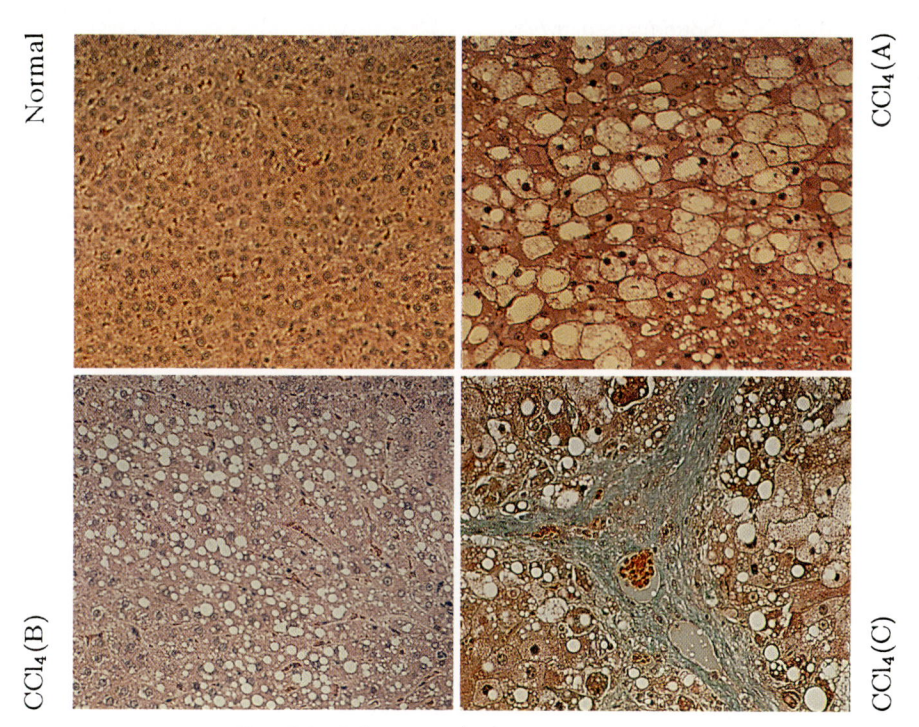

Fig. 5.2 Microscopic image (rat liver)

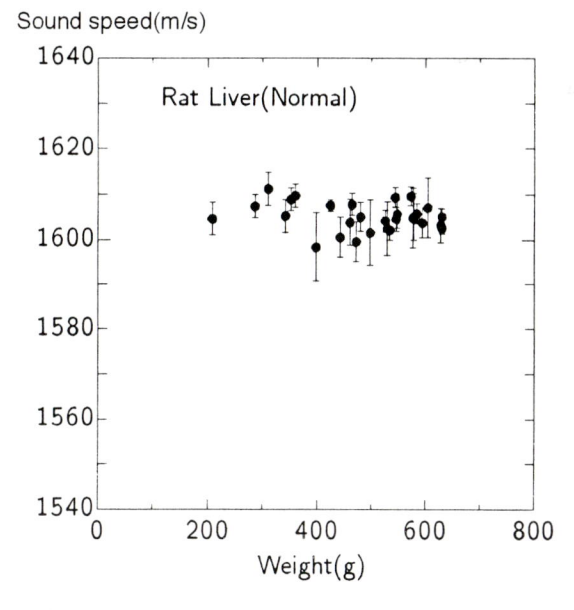

Fig. 5.3 Sound speed of normal rat liver versus body weight

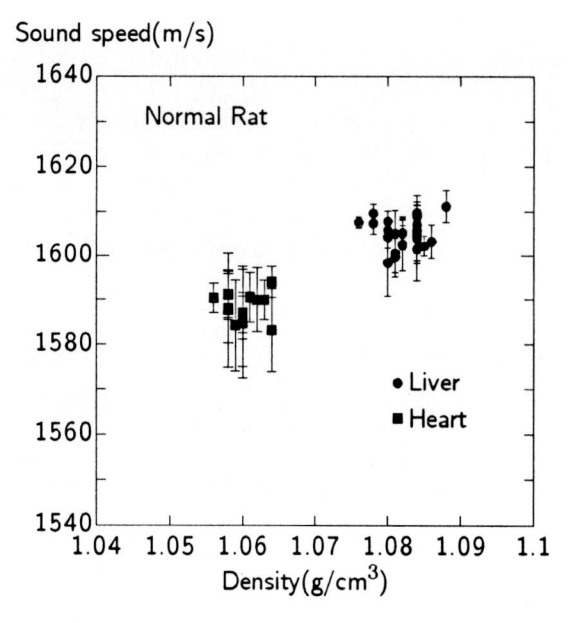

Fig. 5.4 Sound speed versus density of normal rat liver and heart tissues

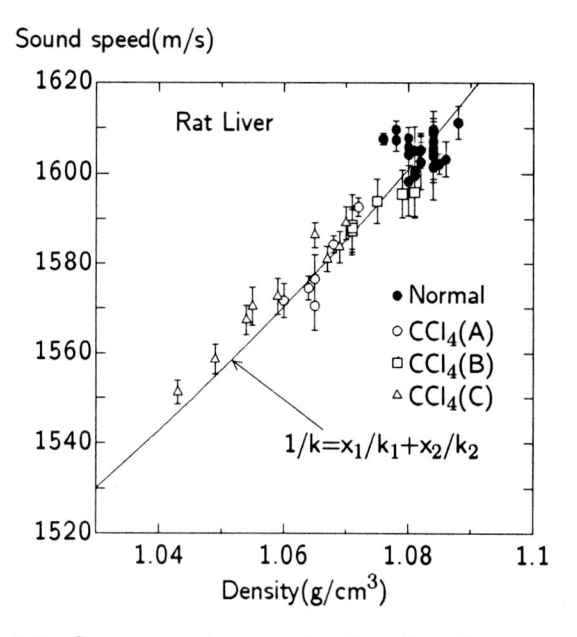

Fig. 5.5 Sound speed versus density of rat liver tissues

5. Non-Contact Measurement of Sound Speed

microscopic view of liver tissue specimens of a group $CCl_4(B)$ rats, we observed many fat droplets in the cells, but the structure of the cell appeared not to change.

The group of $CCl_4(C)$ rats was considered to be cirrhotic liver with fatty degeneration. A stained microscopic section is also shown in Fig. 5.2. The liver tissues of this group felt stiffer than the normal tissues. However, the speed of sound and density of the liver tissues of this group were lower than those of the normal tissues. Hence the bulk modulus was smaller; that is, these tissues were 'softer' than normal tissues.

5.4 The Results at View Point of Fat Content

These results suggest that the decrease of the sound speed and density are caused by fatty degeneration. Assuming a mixture of normal liver and fatty tissue, a theoretical curve can be obtained using an immiscible liquid model. A mixture of two immiscible liquids designated by subscripts 1 and 2, can be described by the equation[6]

$$\frac{1}{\rho c^2} = \frac{x_1}{\rho_1 c_1^2} + \frac{x_2}{\rho_2 c_2^2} \tag{5.3}$$

$$\rho = \rho_1 x_1 + \rho_2 x_2 \tag{5.4}$$

where x_1 and x_2 represent a volume fraction of each component.

$$x_1 + x_2 = 1 \tag{5.5}$$

The average sound speed c_1 and the density ρ_1 of liver tissue of a normal rat are 1604.4 m/s and 1.082 g/cm^3, respectively. The sound speed of fat tissue was measured using the specimen material around the rat kidney. The sound speed c_2 and the density ρ_2 of fat tissue are 1421.2 m/s and 0.908 g/cm^3, respectively.

The solid line in Fig.5.5 shows the calculated relationship between the speed of sound and density. Measurement results of normal and the $CCl_4(A)$ and the $CCl_4(B)$ liver tissues agree well with the calculated results. But the results of the $CCl_4(C)$, which correspond to open triangles, are slightly larger. Thus, the effect of the fibrosis can be shown. There are many reasons for the change of sound speed and density, i.e., it is not clear that the sound speed is related to the degree of fibrosis. For 3.5 MHz ultrasound, the contribution of the collagen to the sound speed may be small, nevertheless the sense of touch of cirrhotic livers is stiffer than that of normal tissues.

Chemical measurement of the fat content was not performed, but the fat content is estimated from the image of the microscopic sections using image processing. This estimated result agrees well with the estimated values from the measured sound speed and density of the rat liver. Thus, the main reason of the change of sound speed is fatty degeneration.

Further work using cirrhotic livers without fatty degeneration is required.

5.5 Conclusion

We have been measuring other specimens. For example, the rat heart, kidney, human liver, and heart. Here we presented our non-contact measurement system and method. For speed of sound, less than 0.2% accuracy is required to distinguish between normal and diseased states. From these measurements, the sound speed of the normal tissue varied minimally and individual differences were less than 0.5% or much less. The relation between sound speed change and physical structure change in liver tissues of diseased rats is becoming clearer.

References

1. H. Hachiya, S. Ohtsuki, M. Tanaka and F. Dunn: "Determination of Sound Speed in Biological Tissues Based on Frequency Analysis of Pulse Response," J. Acoust. Soc. Am. **88** (1992) 1679.

2. H. Hachiya, S. Ohtsuki and M. Tanaka: "Relationship Between Speed of Sound in and Density of Normal and Diseased Rat Livers," Jpn. J. Appl. Phys. **33** (1994) 3130.

3. H. Hachiya, N. Imada, S. Ohtsuki and M. Tanaka: "Relation between Sound Speed and Density of Rat Livers Measured by Non-Deformable Technique Based on Frequency–Time Analysis of Pulse Response," Jpn. J. Med. Ultrasonics. **19** (1992) 633.

4. E. K. McLean, A. E. M. McLean and P. M. Sutton: "An improved method for producing cirrhosis of the liver in rats by simultaneous administration of carbon tetrachloride and phenobarbitone," Br. J. Exp. Pathol. **50** (1969) 502.

5. R. A. Phillips, S. D. Van Slyke, P. B. Hamilton, V. P. Dole, K. Emerson and R. M. Archibald: "Measurement of specific gravities of whole blood and plasma by standard copper sulfate solutions," J. Biol. Chem. **183** (1950) 305.

6. P. L. Chambrè : "Speed of a Plane Wave in a Gross Mixture," J. Acoust. Soc. Am . **26** (1954) 329.

5. Non-Contact Measurement of Sound Speed 71

Discussion

DUNN: We'll now have discussion of Dr. Hachiya's paper. Are there any questions? Dr. Thijssen.

THIJSSEN: You have shown that your reproducibility or specimen variability is on the order of a few meters per second, and then you say this is about the same as the 3 m/s precision of our method. Could you explain a little bit how this measurement precision has been obtained? Is that related to the sampling rate? What other parameters are important?

HACHIYA: To estimate the accuracy of the measurement, I measured a sample of silicone rubber and a sample of acrylic resin many times. This measurement method can measure the thickness d and the sound speed c simultaneously. So the estimation results of the thickness d of the acrylic resin or the silicone rubber was compared with the sequence measurement with a micrometer. Next, using a sample of fresh tissue, we measured the sound speed of the tissue many times. In this way we estimated the accuracy of this measurement.

INOUE: Dr. Hachiya, you have shown in one of your slides that your measurement was volume percent, and the specific density was slightly different. So with the specific density applied, it becomes weight percent?

HACHIYA: Yes, that's correct.

INOUE: And what was the range of the sizes of the fat droplets?

HACHIYA: These are the fat droplet sizes. This dimension is 0.4 mm. Therefore, the size of the fat droplet is on the order of several tens of microns in diameter. So relative to 3.5 MHz being used, this tissue looks homogeneous.

INOUE: Yes, I think so.

HACHIYA: In this case, for the 3.5 MHz, this tissue looks homogeneous, not totally uniform, but it is homogeneous. This is about half the width of the beam.

SASAKI: I'm from Tohoku University. I have two questions, Dr. Hachiya. First, how long does it take to make one measurement? You use fresh tissue specimens after excising the tissue from the biological system. From the moment it is excised, degeneration processes start. Have you considered the degenerative change because of excision of the tissue?

HACHIYA: The measurement was done within 30 minutes after excision of the tissue, in the case of rat liver. The study was completed within 30 minutes after the excision. After excision and immersion in the saline, we looked at the degenerative process. Sound speed change was within 0.1 % for 100 minutes. After that, this changes further. So if the measurement was completed within 30 minutes, there was no significant influence on the measurement due to the degenerative process after excision. Thank you very much.

SASAKI: Another question. In case of liver–and this is the same with the other organs–but in liver in particular there are many vessels, arteries, and veins. So it is very rich in vessels; it's a very well-perfused tissue. If in the tissue you study there are many vessels, this must have some impact on the speed of the sound, which we have experienced as well. Did you look at the influence of the circulation on the lipid deposition?

HACHIYA: In this measurement, 4 by 4 mm was the range in which the measurement was taken. And that's where we have placed the tissue. We can also observe internal echoes, which were sometimes greater compared to the reflection from the surface, and in such cases that part was excluded. So with a 4 by 4 mm size, we excluded that kind of area, because that suggests there was a very big structure inside. Thank you.

KUSHIBIKI: You made comparison measurements between normal liver and fatty liver in terms of the average values?

HACHIYA: Yes, in rats.

KUSHIBIKI: The percentage of fat to liver tissue,–how much was it? What was the fat percentage?

HACHIYA: According to acoustic estimation, 20 % maximum. In such case, compared to the measurements done by Dr. Inoue, for example, I believe that the speed we have shown is much slower, and it seems that the density is very low. It's very close to the density of the fat.

KUSHIBIKI: Your measurements of the fatty liver compared to those by Dr. Inoue look very small in terms of speed and density. I believe the density is low–0.9 something–but if the fat represents only 20 % of the tissue, then the measurement of the tissue density looks too small. Any comment, Dr. Inoue?

INOUE: Well, there are differences. First of all, it's a difference between rat versus pig, maybe. Different animals; species difference. But in our case, it was the synthesized value. Probably the fat that was in the liver is contributing to a certain amount of error. The tendency is correct, although it was the synthesized value.

KUSHIBIKI: Did you look at the speed of sound and the density of the fat itself, and is it different than my results?

INOUE: In the fat itself, the density in the liver is 1500 or something, and the speed of sound is 1580 m/s, and in fat 1410 m/s. So it's lower in the fat by quite a bit.

HACHIYA: In response to the question by Dr. Kushibiki, in our case 1604 m/s was the measurement in the liver, and then 1420 m/s for the fat. In other words, there is no discrepancy.

KUSHIBIKI: Yes, but the 1420 m/s for the fatty liver, in which only 20 % fat is included.

HACHIYA: No, that is the fat tissue itself.

KUSHIBIKI: Then that's very close. I understand now.

DUNN: Any further question? If not, we have concluded session I.

Part II

CLINICAL APPLICATIONS

Chapter 6

Multiparameter Ultrasonic Tissue Characterization and Image Processing: from Experiment to Clinical Application

6.1 Introduction

We started in the early 1980s with some serious efforts in tissue characterization, and have worked on the effects of beam characteristics on measurement of acoustic parameters, and on speckle characteristics of B-mode images. The methods and techniques we developed have been applied in clinical studies on diffuse liver disease and eye tumors. Two of my students finished their Ph.D. in 1990 [1,2], and after that date we proceeded with image processing and acoustic microscopy. Image processing was meant to start the next phase in tissue characterization. Since, we had developed the methods and the tools, and shown in clinical practice that it could work [3-7], we wanted to try and find methods to produce convincing images from these results. And at the same time, we started with acoustic microscopy in the hope of solving basic problems we had encountered while doing the clinical studies. We not only had obtained some fine results, but also we got some new interesting questions from these.

In the mean time, Van der Steen proceeded in writing a thesis on acoustic microscopy that was finished early this year [8], he worked primarily on liver tissues [9,10]. De Korte used the acoustic microscope for measuring the acoustic

---
*J.M. Thijssen, Ph.D.

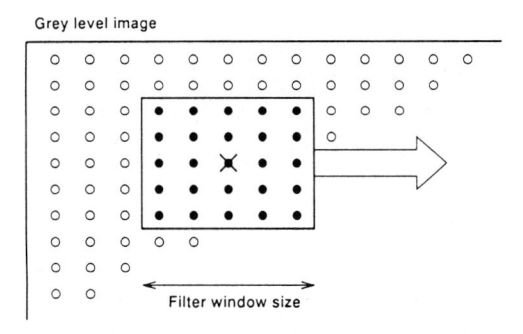

Fig. 6.1 Sliding window technique

characteristics of eye tissues [11-13]. Further recent activities were devoted to parametric imaging and image processing. The results of these studies were summarized in the Ph.D. thesis by Verhoeven [14] and in many publications which are discussed in the following Sections of this paper.

6.2 Parametric Imaging

The distinction between parametric imaging and image processing is kind of artificial. Nevertheless, we want to distinguish these two methods, these two means of processing. The major goal of processing the echographic data is to display all the information that is present in echograms. This will serve to improve the detection of diseases, and if we have found the disease, for instance a tumor or some kind of diffuse change in tissue, to subsequently differentiate and say which kind of disease, which kind of tumor we have detected. This not only applies to the eye, of course, but to other organs as well. We have been developing methods by using liver, and also simulated data, as a model while generally having ophthalmological applications in mind. The methods we have been using are parametric imaging and image processing, with the final goal to try to segment the image and determine which part of the organ or body is diseased. And I must stress that we did not reach the phase that we have the computer doing the segmentation, and this is really a topic that is challenging for the future.

Parametric imaging is based on analyzing the data, and we do that in overlapping windows, i.e. the window is as it were sliding over an image, like is sketched in Fig. 6.1. We can do both an analysis of the radio-frequency signals, or of the video image itself. After the analysis we have to perform a pixel-wise encoding to create a new (parametric) image. This explains in nutshell what we mean by sliding window. Each pixel of an image, except for the border which we cannot process in this way, surrounded by this window will get a value which is determined for the whole window. It will be obvious, that if one wants to have

6. Multiparameter TC and Image Processing

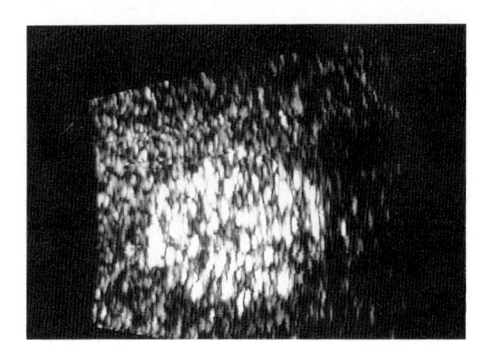

Fig. 6.2 B-mode image of scattering phantom, illustrating "speckle". Central part: lesion containing larger size scatters than in surrounding medium

a very precise parameter estimation, a large window is needed, and if a large window is used, resolution is low. So, it is always a trade-off between precision and resolution. Furthermore, the employed window size will be different for each different parameter. For instance, to estimate the attenuation image, a different window has to be used than for backscattering, and vice versa. The choice is depending on the precision of the estimation algorithm.

6.2.1 Acoustospectrograhpic Imaging

When we talk about parametric imaging, let us say performed by MRI, i.e. magnetic resonance spectroscopic imaging (MRS), we could state that we are performing ultrasonic spectroscopic imaging, because what we basically do is estimating the frequency dependence of the attenuation coefficient and of the backscattering coefficient.

The B-mode image in Fig. 6. 2 was made from a tissue-mimicking phantom, which we constructed by using gelatin-based material where we had small silica gel scatters in the base material and larger size scatters in a cylindrical central zone. This larger size caused a little bit more backscattering, so that is why we see a lesion with a positive contrast in this B-mode image. To go quickly through the acoustic spectrographic imaging, when we go through the tissue in depth, we will get an attenuation of the spectrum corresponding to the transmitted ultrasound pulse, which is higher the higher the frequency we have (Fig. 6.3, left). So high frequencies attenuate a lot, and low frequencies attenuate much less. If we normalize the spectrum at each depth to its maximum, we obtain spectra as shown in Fig. 6.3, right part. One can observe that after the normalization the spectrum apparently shifts to lower frequencies due to the attenuation. This is what we call the spectral shift attenuation estimation method. What is shown in Fig. 6.4 is an example of the application of this method where we imaged the instantaneous frequency as a function of depth, which is a measure of the

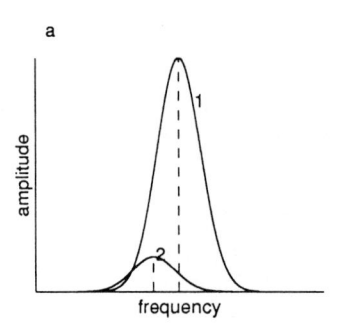

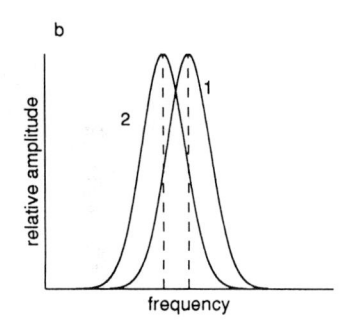

Fig. 6.3 Left: spectra corresponding to received rf-echoes received from increasing depth (2 vs. 1). Right: same spectra after normalization of maximum value [25]

attenuation coefficient. The color of the image changes from blue to yellow-red. It means that the frequency has shifted downwards. But you can also see that in the area where we had the bright lesion in Fig. 6.2, we now see a higher attenuation, as well. This means that the larger structures in that part of the tissue-mimicking phantom cause a higher attenuation. This proves that we can make a local estimate of the attenuation as function of frequency, basically, because this is a frequency-shift method. And we can easily calculate from the frequency shift the attenuation coefficient, although we do not need to do that. We just want to make an image highlighting an interesting region of the image. So, we have shown here that, in principle one can make images of the attenuation coefficient and show a lesion.

I should emphasize that all this kind of work has to be done after we have made the data independent of the depth to a certain extent. So it means in this case, we process the radiofrequency data in the propagation direction along the A-lines and correct for the beam diffraction, and then after we have estimated the attenuation coefficient we correct for the frequency-dependent attenuation coefficient [3,5]. Then we have data which are suitable to estimate the backscattering as a function of frequency. In Fig. 6.5 the backscatter intensity, relative in decibels, is shown as a function of frequency in the range of zero to 10 MHz, which is the scattering structures. In this case, these curves correspond to spherical particles. So this information is purely derived from physics, and we can calculate the backscatter intensity for spherical particles as a function of size.

Now if we worked with a transducer intensity for spherical particles as function of size. Now if we worked with a transducer, let us say, of 7 MHz, then we have a range of frequencies that we can use from 4 to 10 MHz, for instance. Within such a range, these curves can easily be approximated by straight lines. The slope of these straight line fits is uniquely related to the size of the particles. So a positive slope at one curve means a 20 micrometer particle, whereas a negative slope of another curve means a particle size of over 50 micrometers. So one can assess the slope of the linear fits to measured backscattering spectra, and then

6. Multiparameter TC and Image Processing

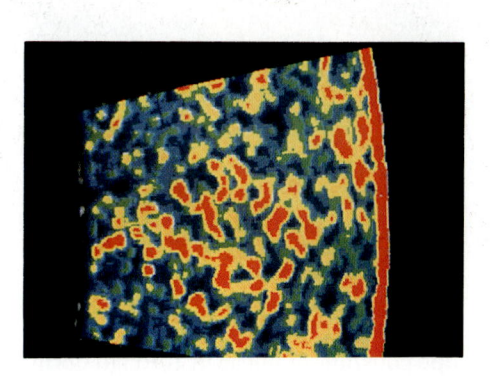

Fig. 6.4 Attenuation image corresponding to Fig. 6.2, obtained from instantaneous frequency decay [2]

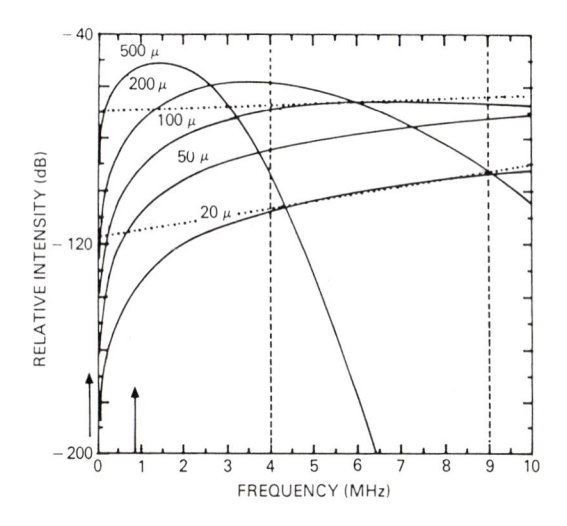

Fig. 6.5 Theoretical curves of normalized backscattering coefficient vs. frequency for a Gaussian spatial correlation model with various range constants, as indicated. Linear fit within the -6 dB frequency range of transducer response (4-9 MHz) is indicated

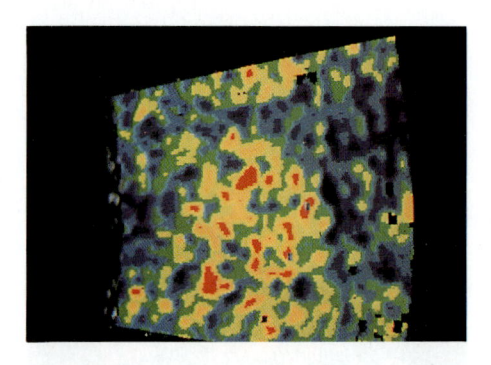

Fig. 6.6 Image of slope constant of linear fit through normalized backscattering spectrum, rf-data of Fig. 6.2 were employed

relate it to the size of the particles.

Now what we have done in Fig. 6.6 is that we have plotted the slope of the linear fit through the backscatter spectra, locally estimated in windows, as a function of position like before for the attenuation. It can be appreciated that indeed in the area where we had inserted in the phantom larger particles, we get on average a lower slope, as we would expect from Fig. 6.5. So, we have the lower values coded with red, as we did before with the attenuation coefficient, and here we get a nice color contrast for the larger size of the backscattering particles. Again a proof is given, although with a tissue mimicking phantom, that by locally estimating acoustospectographic parameters, you can find the lesion. Even lesions without contrast in the B-mode image can be made visible with this technique.

6.2.2 Texture Parameter Imaging

We have to introduce here the word speckle, which may be defined as the smallest "information grain" in an echographic image of a scattering medium. It is caused by the statistical interference of backscattered echoes at reception by the transducer. When using the sliding window technique for imaging of texture parameters, the shape of the window has to be adapted to the dimensions of the speckle. Basically, in the axial direction the resolution of the echogram is much better in general than in the lateral direction, so the speckles are laterally elongated (Fig. 6.2). It means that when using windows, one has to look at the shape of the point-spread function (PSF) of the employed transducer. That is the adequate reference to use in the focal zone of the employed transducer. But there is another important characteristic of speckle: if one is estimating speckle characteristics in B-mode images, it appears that speckle size is not directly related everywhere to the PSF. So, for instance near to the transducer, the speckle

6. Multiparameter TC and Image Processing

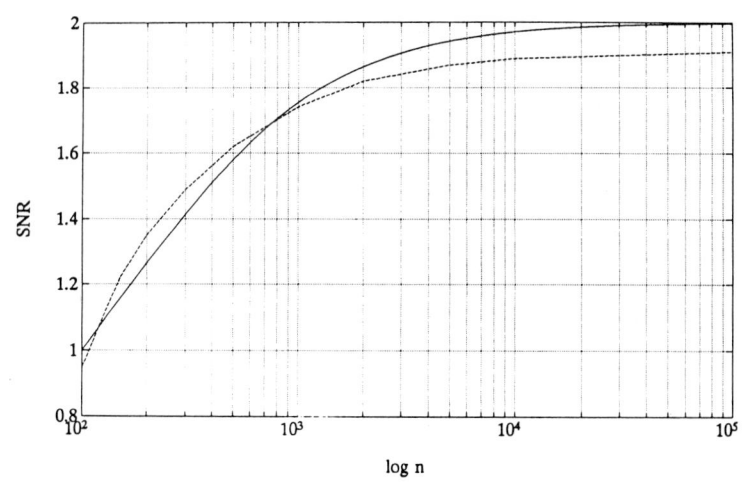

Fig. 6.7 Signal-to-noise ratio (SNR) vs. logarithm of scatter number density N. Dotted line: SNR of envelope (gray level), as simulated by Oosterveld et al. [15]; drawn line: SNR of envelope obtained from Eq.(6.1)

in the image is very fine, although the point-spread function is very wide there [15]. This property of speckle induces the need to adapt the size of the window to the depth into the tissue, i.e. to the distance to the transducer surface. In summary, the shape of the window is depending on the dimensions of the PSF, while the size of the window depends on depth and on the parameter that is being calculated.

In texture parameter imaging (TPI) different parameters which are related to the image texture are locally estimated. Texture can be defined as the spatial distribution of gray levels in an image and which is related to the structural characteristics (histology) of tissue. The first texture parameter is the signal-to-noise ratio of the gray levels, the second one is the L2 mean, and the third one is the fractal dimension.

Let us start with the signal-to-noise ratio (SNR). It is simply the mean gray level value over the standard deviation. Both of these parameters are obtained from the local gray level histogram, where the "local" means within each window. About 10 years ago we performed a rather extensive simulation study with the slow computers we had available at that time. Oosterveld et al. [15] produced basically the dotted curve that is shown in Fig. 6.7, where the SNR is plotted vs. the number of scatters per cm^{-3}, (from 100 to 10,000). If we increase this "number density", calculate B-mode images and then assess the SNR, we see that this parameter increases by about a factor of two, from about one to about two (1.91). When this high scatter density limit is approximately reached the gray level histogram approaches a Rayleigh probability density function for which the SNR=1.91. The speckle is then said to be "fully developed". This was done for a transducer of 3.5 MHz. It means that in the range from 100 to about

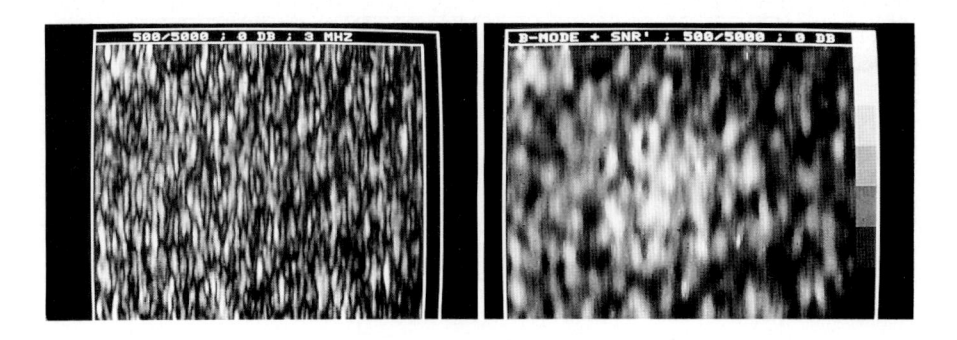

Fig. 6.8 Left: simulated image of echogram with isoechoic lesion (number density 500 cm^{-3} vs. 5000 cm^{-3} of surrounding medium). Right: SNR image calculated from these data [26]

5,000 scatters per cm^{-3} we can distinguish the texture related to the scattering media by looking at the SNR. It is a parameter that distinguishes tissues with different scatter densities independent of the local gray level (scattering strength) differences! Now there is also a formula that has been derived by Jakeman [16], who is working in the radar field. There, similar problems are tackled as in the ultrasound field. He showed that for the SNR of the intensity, which is the square of the amplitude this relatively simple formula, is obtained for sub-Rayleigh statistics:

$$SNR_I = \left(1 + \frac{<a^4>}{N<a^2>^2}\right)^{-\frac{1}{2}} \quad (6.1)$$

where a is the scatter strength and N is the number of scatters in the resolution volume.

You see in Eq. (6.1) that the ensemble average of the scatter strength, so it means that the different scatters do not need to have the same scattering strengths individually. If you plot this formula, you get the drawn line in Fig. 6.7 (the data to construct this curve were multiplied by two, because for the SNR of the intensity the Rayleigh limit equals one). So with this theoretical work, the same range from about one to two for the signal noise ratio is found as with the simulations. In other words, if one would plot the SNR of the intensity, the same potential to perform a differentiation between tissues is achieved. Fig. 6.8 shows an example of an image we have simulated, where in the center there is an area where the scatter density is lower -i.e. 500 as compared to 5,000 cm^{-3} in the environment of the lesion. Because it is a computer program, we can adapt the scattering strength within the lesion in such a way that it has exactly zero contrast. And now we see the limitation of the human brain. We cannot perceive a difference in the second order characteristics of B-mode images, so we do not see the lesion here. However, we know it is just there and the size of the speckles

6. Multiparameter TC and Image Processing

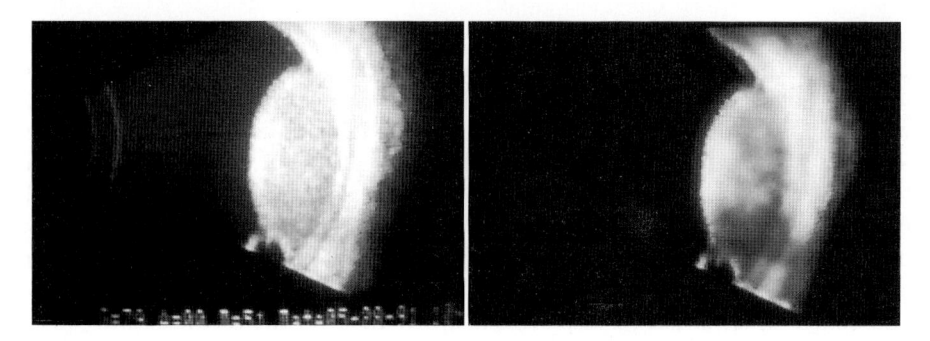

Fig. 6.9 Left: B-mode image of backwall of human eye, showing a large tumor. Right: same image, obtained by L2-mean filter

within the lesion are just a little bit different from the surrounding tissue. Now, if we make an image of the SNR, and in this case we have combined it simply with the original B-mode image by taking the square root of the sum of the squared magnitudes of gray level and SNR, one can perceive exactly where the lesion is. So by using this relatively simple texture feature, the signal-to-noise ratio, we can produce an image which shows isoechoic lesions.

6.3 Image Processing

Image processing is often related to image filtering, so one may find in handbooks the mean filter, median filter, L2 mean filter, and so on, under the heading of imaging processing, and of filtering in other books. I do not know exactly what the difference would be, but anyway Fig. 6.9 a parametric image is calculated, where as an example the square of the echo amplitude is calculated for each pixel and then the average over the whole window is taken, i.e. the L2-mean is depicted.

L2-mean: definition

$$M_{L2} = \frac{1}{nm} \sum_{i=1}^{n} \sum_{j=1}^{m} (P_{ij})^2 \qquad (6.2)$$

But one may also consider this operation to be a low-pass filter with a certain kernel. If you do that, you can also take it one step further, and say, well, we make this filter adaptive:

$$I_{ij} = k_{ij} P_{ij} + (1 - k_{ij}) f(P_{ij}) \qquad (6.3)$$

where, $k_{ij} = g(a, P_{ij}) k_{ij} \in (0, 1)$, g = a continuous function, a = "agressiveness" parameter, f = filter operator (e.g. L2-mean, Eq.(6.2)), P_{ij} = pixel(gray level)value

This formula shows that the filter characteristics are to be changed at each position within the image, depending on a certain characteristic of the image itself. That is what we call adaptive filtering. It becomes clear from Eq. 6.3 what happens. If k is zero, we change the pixel in the center of the window according to the filter (e.g. L2-mean, Eq. 6.2). But if k equals one, we leave the pixel as it is. For instance, if we have some kind of edge detection in this k factor, we can say that if we find a contour in a B-mode image, we do not use this L2 mean filter. But if we do not find a contour, we have to replace the central pixel value by the L2-mean. This means that we get preservation of the anatomical information by using this adaptive mode of the L2 mean filter, or in general, of the texture parameter filter.

Figure 6.9 shows an example. The left picture is a conventional B-mode image of a human eye containing a choroidal tumor (melanoma). The right picture in Fig. 6.9 shows the result of the adaptive L2-mean filter. This latter produces a considerable speckle reduction, whereas the contours which are related to the anatomy are almost completely preserved.

Now why did we choose the L2 mean for creating this adaptive filter? It can be proven that, if we have fully developed speckle, i.e. a high scatter number density, then the SNR of the image becomes 1.91 and the optimal filter should be the L2 mean filter [17,18]. This property can be simply derived from the Rayleigh probability distribution function, which applies to fully developed speckle. In the eye, we can always see the tumor easily with ultrasound, there is no problem, but Fig. 6.9 just serves as an illustration of the method.

The two texture parameters discussed so far (SNR, L2-mean) have been taken from the gray level histogram. One could also look at the speckle pattern, i.e. the second order statistics of the gray level distribution (i.e. the texture). One of the methods we have tried and which proved to be successful is the "fractal" analysis [19]. Fractal analysis is well known for optical images, but not so well known, I think, for acoustical images. While this method is not proven to be the most appropriate for applying to B-mode images, it worked well for the question we had. We wanted to have a second-order parameter which was easily calculable from relatively small windows. A potential problem with second-order statistics is that large windows might be needed to achieve precise estimates of the parameters. The basic principle as shown here is based on the scale-space filtering approach [20] where the original intensity scale image, or B-mode image, is a 2D convolved with a kernel:

$$I = P * *G(\sigma) \qquad (6.4)$$

where $G(\sigma)$ = 2-D Gaussian $(0, \sigma)$ kernel, and * means convolution

The optimal kernel can be shown to be a Gaussian function [21]. The basic result of such a kernel is a low pass filtering of the original gray scale image. The area of the 2-D intensity surface, $A(\sigma)$, corresponding to the thus derived new gray scale image follows from:

6. Multiparameter TC and Image Processing

Scale Space Filtering
smoothing

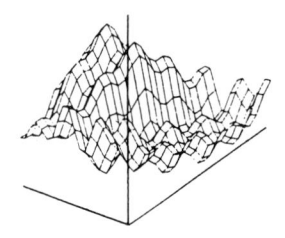

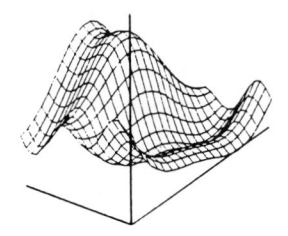

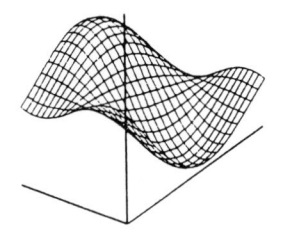

Fig. 6.10 Plot of gray level (vertical) vs. position in B-mode image. Left: original; middle and right: smoothing due to increasing σ of Gaussian kernel (Eq.(6.4)) [19]

$$A(\sigma) = \int\int (1 + I_x^2 + I_y^2)^{1/2} dx dy \qquad (6.5)$$

where I_x, and I_y are the first derivatives of $I(x,y)$ with respect to x and y, respectively.

The scale space filtering is illustrated with Fig. 6.10, where the intensity of the B-mode image is plotted vertically and the effect of increasing σ of the Gaussian is shown from left to right.

Now there is some very old theory of the 19th century or beginning of this century from which the Minkovsky dimension can be derived. It can be shown [22] that for the convolution with this kernel the Minkovski dimension M is proportional to the slope of the curve relating the logaritm of the logarithm of the intensity surface to the logarithm of the σ of the Gaussian:

$$M = T - \frac{\ln A(\sigma)}{\ln(\sigma)} \qquad (6.6)$$

where T = topological dimension of B-mode image(i.e. $T = 2$)

The next step is then to plot an image of the Minkovsky dimension. The left part of Fig. 6.11 shows the image as displayed in Fig. 6.8, i.e. with an isoechoic low-number density lesion, which is not visible because we made the contrast zero. If we then compare this with the Minkovsky dimension image, i.e. the so-called fractal image (Fig. 6.11, right), it appears clearly where the lesion is. And in this case, we have shown here in inverse gray level code a parameter related to the size of the speckles.

We have till here discussed several parameters, derived from the attenuation, from the backscattering, and some from the image texture. We also did efforts, when confining ourselves to three parameters to make a color-coded image. It

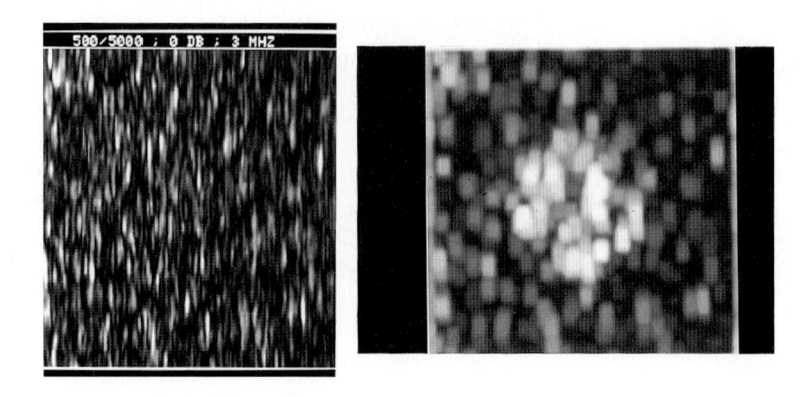

Fig. 6.11 Left: identical to Fig. 6.8 left part. Right: fractal (Minkovsky) dimension image [19]

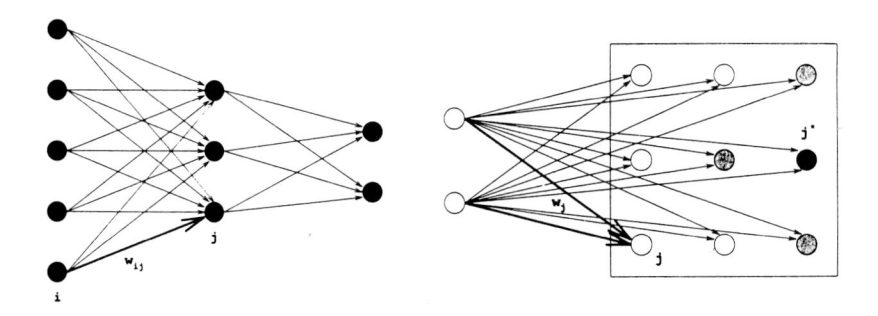

Fig. 6.12 Left: scheme of feedforward backpropagation artificial neural network (ANN). Right: self organizing feature map ANN [23]

appeared not very easy to find a good coding of parameters in colors. Because if you start using the hue (coded red to green,), the saturation (from white to full color), and luminance (from white to black), in principle three independent axes of the color space are available. These images did, however, not look nice to the human observer. The problem is due to the restrictions of the red and green and blue color channels of the TV monitor. In optical systems, it would have been much more adequate to image three parameters. Therefore, we stopped trying multiparameter coding in color.

We have concentrated more on finding methods to reduce multiparameter decision-making to single-parameter decision-making, and the classical way of doing that is the Fisher's discriminant analysis. Otherwise you have to take, for instance, a K -nearest- neighbor method. But finally, we also tried to improve on that by using artificial neural networks (ANN) for decision making. We tried out a feedforward ANN with a back propagation learning rule (Fig. 6. 12, left part), which is a supervised method. So during the training, the deviation of the output from the true value is minimized by adjusting the weight constants in the network. Another type of ANN is a self-organizing feature map (Fig. 6. 12, right part), which runs completely unsupervised, but proved to be less efficient than the back propagation method in the end [23]. However, both ANN's proved to yield better results than the discriminant analysis. So, the ANN's are maybe better models to achieve a reasonable decision making in case of a limited data set. Although it is not explicitly stated in the foregoing sections, I would like to stress that the methods are based on specific models of the tissues. The wave propagation model is supposed to be linear, so beam diffraction, attenuation and backscattering are separable. The tissue model is further specified by random plus structural scattering. Furthermore, the model is based on the clinical experience we have gathered during the years. It depends on the kind of problem–for instance, what organ are you working in, or what kinds of diseases are you trying to differentiate, which choice of parameters will be made. All in all, we have at least nine or ten well-behaving parameters [3,6,24], which could be related to acoustic tissue models. Depending on the number of proven clinical cases, this number has to be reduced and it is necessary to select for each clinical question the most optimal ones. That is at least the strategy we have followed till now. We have also concluded that a single parameter is never sufficient for clinical diagnosis. So we need to have a multiparameter approach where the parameters are based on different mechanisms of interaction of ultrasound with tissues. Finally, we need to develop some kind of multiparameter imaging technique which is not slowing down the whole diagnostic process. If we want to have real-time behavior, we should have a trained system in terms of neural networks. The ANN is trained with a great number of data, and then when it is tuned it can make an immediate decision. That is the strength, of course, of neural networks. So one could imagine a kind of ANN that has been trained to find tumors in the liver and another one fit to detect diffuse liver diseases.

As stated before, image processing and prametric imaging are not really much different, but I would like to add that there are some other possibly useful methods in image processing books and publication. These are encouraging enough to state that, maybe, the processing of the final multiparameter image could improve the diagnostic potentials, by a kind of "finishing touch". This could be also important for both the clinical- and the commercial acceptance of the methods we have been working on.

I hope that this paper will at least stimulate the young scientists in this area, and that those scientists who have been in the field for a longer time find some satisfaction in the conclusion that there is still progress.

References

1. Oosterveld BJ (1990) On the quantitative analysis of ultrasound signals

with applications to diffuse liver disease. Ph.D. Thesis, Nijimegen University.

2. Romijn RL (1990) On the quantitative analysis of ultrasound signals and application to intraocular melanomas. Ph.D. Thesis, Nijmegen University.

3. Romijn RL, Thijssen JM, Oosterveld BJ, Verbeek AM (1991) Ultrasonic differentiation of intraocular melanomas: parameters and estimation methods. Ultrasonic Imaging 13: 27-55

4. Thijssen JM, Oosterveld BJ (1990) Texture in tissue echograms. Speckle or information? J Ultrasound Med 9: 215-229

5. Oosterveld BJ, Thijssen JM, Hartman PC, Romijn RL, Rosenbusch GJE (1991) Ultrasound attenuation and texture analysis of diffuse liver disease: methods and preliminary results. Phys Med Biol 36, No. 8: 1039-1064

6. Oosterveld BJ, Thijssen JM, Hartman PC, Romijn RL, Rosenbusch GJE (1993) Detection of diffuse liver disease by quantitative echography. Ultrasound Med Biol 19: 21-25

7. Hartman PC, Oosterveld BJ, Thijssen JM, Rosenbusch GJE, Berg Jvan den (1993) Detection and differentiation of diffuse liver disease by quantitative echography. A retrospective assessment. Invest Radiol 28: 1-6

8. van der Steen AFW (1994) Acoustic microscopy: an in vitro tool for development of ultrasonic tissue characterization. Ph.D. Thesis, Nijmegen University.

9. van der Steen AFW, Thijssen JM, van der Laak JAWM, Ebben GP, de Wilde PC (1994) A new method to correlate acoustic spectroscopic microscopy (30MHz) and light microscopy. Microscopy 175: 21-33

10. Steen AFW van der, Thijssen JM, Laak JAWM van der et al. (1994) Correlation of histology and acoustic parameters of liver tissue on a microscopic scale. Ultrasound Med Biol 20: 177-186

11. de Kort CL, van der Steen AFW, Thijssen JM (1994) Acoustic velocity and attenuation of eye tissues at 20 MHz. Ultrasound Med Biol 20: 471-480

12. de Kort CL, van der Steen AFW, Thijssen JM et al. (1994). Relation between local acoustic parameters and protein distribution in human and porcine lenses. Exp Eye Res 59: in press.

13. van der Steen AFW, de Korte CL, Thijssen JM (1994) Ultrasonic spectroscopy of the porcine eye lens. Ultrasound Med Biol 20: in press.

14. Verhoeven JTM (1994) Improvement of echographic image quality by data analysis and processing. Ph.D. Thesis, Nijmegen University.

15. Oosterveld BJ, Thijessen JM, Verhoef WA (1985) Texture of B-mode echograms, 3-D simulations and experiments of the effects of diffraction and scatter density. Ultrasonic Imag 7: 142-160

16. Jakeman E (1984) Speckle statistics with a small number of scatters. Opt Eng 23: 453-461

17. Kotropoulos C, Pitas I (1992) Optimum nonlinear signal detection and estimation in the presence of ultrasonic speckle. Ultrasonic Imag 14: 249-275

18. Verhoeven JTM, Thijssen JM (1993) Improvement of lesion detectability by speckle reduction filtering. Ultrasonic Imag 15: 181-204

19. Verhoeven JTM, Thijssen JM (1993) Potential of fractal analysis for lesion detection in echographic images. Ultrason Imag 15: 304-323

20. Witkin AP(1983) Scale space filtering. In: Proceeding Joint Conference on Artificial Intelligence. Karlsruhe, pp. 329-332

21. Baband J, Witkin AP, Baudin M, Duda RO (1986) Uniqueness of Gaussian kernel for scale space filtering. IEEE Trans Patt Anal Mach Intell 10: 704-707

22. Msigmann U (1989) Texture analysis, fractal and scale space filtering. In: Proceedings 6th Scandinavian Conference on Image analysis. Oslo, pp. 987-994

23. klein Gebbink MS, Verhoeven JTM, Thijssen JM, Schouten TE (1993) Application of neural networks for the classification of diffuse liver disease by quantitative echography. Ultrasonic Imaging 15: 205-217

24. Thijssen JM, Oosterveld BJ, Hartman PC, Rosenbusch GJE (1993) Correlations between acoustic and texture parameters from RF and B-mode liver echograms. Ultrasound Med Biol 19: 13-20

25. Thijssen JM, Verbeek AM, Romijn RL, de Wolf-Rouendaal D, Oosterhuis JA (1991) Echographic differentiation of histological types of intraocular melanoma. Ultrasound Med Biol 17: 127-138

26. Verhoeven JTM, Thijssen JM, Theeuwes AG (1991) Improvement of lesion detection by echographic image processing: signal-to-noise-ratio imaging. Ultrasonic Imaging 13: 238-251

Discussion

OPHIR: Thank you very much, Dr. Thijssen. The paper is now open for questions. Professor Waag.

WAAG: I'd like to thank you for your presentation and ask you for some additional comment about your parametric imaging. In particular, I noticed that you do some signal analysis and then derive from that a parameter, and then you plot that parameter in an image. My request is for you to give us some idea about the intervals in time and in space that you're using to carry out your analyses to get a pixel, and tell us a little bit about the stability of these analyses.

THIJSSEN: OK. Let me try to understand your question. You're relating the methods to the size of the windows we can use to estimate these parameters. I think for the textural parameters, we have about three times the size of a speckle as a window. So it means that if you have a speckle that is five pixels in the depth direction, then it can be about 15 pixels wide in the lateral directions, which is bigger, and then you take something like two to three times that size. So you have a few speckles within one window: otherwise, you don't have a large enough statistical sample of the characteristics of the histogram. So you need that statistical point of view. I think this could equally apply to acoustical parameters. We have not done a systematic study on that, but we use about the same approach. You can find in the literature some publications where, for instance, the estimation of the attenuation coefficient slope parameter is related to window size. There are theoretical formulas for that. Then you can say OK, I want to have a precision of 10 percent of the attenuation coefficient, and you calculate the window size you would need given the bandwidth and frequency. And that would work. And I estimate from my memory that it's about the same. So you need about three times the size of speckle in the area and the depth range where you are to estimate the width and precision on the order of 10 percent to 25 percent. So it's a not very precise estimate you make of the acoustic parameters. Then, if you make an image, one will perceive some color on average, and that's the way it works for visual use of these images. It's semi-quantitative in that sense. Now there is some literature–I think Roman Kuc - if you remember him, who has been working in tissue characterization for many years–and he estimated that for a 3.5 MHz transducer an area of 2 by 2 centimeters, so let's say 4 to 5 square centimeters, would be needed to get accurate estimates of the attenuation coefficient. That's much bigger than we're talking about here, but he was still maybe talking like a physicist and not like an engineer. If that's an answer to your question.

WAAG: Well, yes, that answers it. I think the experience one would infer from your images is that the resolution cell is much smaller than some of these theoretical predictions. I think we get some of that analysis too in our laboratory. We came up with an optimistic scale of about one centimeter, which is many more than three speckles, in these images. Could you tell us how big that object was that you were displaying in your...

THIJSSEN: That was 3 centimeters in diameter. That was a big one. So you have even after the image processing, or after the parametric display, in the axial direction about 20 speckles in there. That's why you get the good impression of a single volume. It's not a misleading image, but it's not maybe a clinically

6. Multiparameter TC and Image Processing

relevant image yet.

QUESTION: I have a follow-up question. If you say it's not a clinical image yet, what do you expect to happen? If the theory says, like Roman Kuc's theory or whoever's theory, says that you need 1 or 2 centimeters, what else can you do? In other words, I thought you implied that it's not yet useful but it will be useful? So what else can you do to push the theory further?

THIJSSEN: I already had one example. That is where we combined the original B-mode image with the SNR information, and we did that in a kind of vector summation. The point is that it may be too simple to do it in this way. That's why we're looking now for more sophisticated methods to combine parameters. And the basic message is that all these images, more or less, have been single parameter images. And that's the improvement we want to get. So you can make smaller-sized windows if you combine more parameters because you enhance the information about the tissue by combining. So for instance, if you have cirrhosis, you can have less acoustic scatters of a higher scattering strength. So you get several of these parameters, which change simultaneously in one direction, so that would enhance it. That's the hope we have.

QUESTION: So the essence of what you're saying is, that if you think that you're going to detect disease with one parameter, forget it. This is basically what you're saying?

THIJSSEN: I think that's one of the reasons we proceeded as we did. A lot of people worked on attenuation or on this or that, and we hope now to combine that knowledge and do it quickly in real-time. What we consider real-time is what we call clinical real-time. And clinical real-time means not in a fraction of a second, but in a second or so. So within one eye blink of the radiologist, a new image is created. That's clinical real-time.

DUNN: Dr. Thijssen, can you tell us a little bit more about the neural network approach. Does it appear to be very promising at all? And if so, what it take to make it useful?

THIJSSEN: Well, I think somebody, maybe even in the audience, will say that neural networks are just a dirty trick invented by some mathematician. But on the other hand, we have about 120 billion of these things in our brain, so it's working. It means that it's a flexible network of small processing units, and the point is that like in the brain, by activating the synapses, which are the weighting factors between the successive neurons, in a practical way, it means by using information that is practical, you can kind of fix a pattern of waves that propagates a certain input to certain output. That's the basic thing. That's how we also learn. If you get a certain stimulus, external effect, then you have a certain reaction. That is a path that has been established by our own experience in our own brains. That's the way it works. The point is that discriminant analysis, which is also a learning system, in general is based on linear or quadratic approaches. But a neural network is to a certain extent unpredictable, there's no theory of how it exactly works. This feed-forward back-propagation network has no firm theory. It's highly non-linear system. So you cannot predict from the description of the network how it will behave in a certain situation. Moreover, it's plastic and as I said it's unpredictable. But if you train it with great number of known cases, and that's what we do all the time in our tissue characterization, - and the same way we do it with the neural network - if we train it with the proper information, then in the end it will give us immediate answers which are correct to a high extent, often higher than these other methods do. And I think that's the promising fact about neural networks. It's unpredictable, but if you train well, it's good.

QUESTION: Thank you. Will we have some practical application from the clinical point of view?

THIJSSEN: Yes, but I think there are some practical problems. What we're doing now in Europe is that we have a European Union contract to establish four institutions where we have the same software to do the analysis of the rf-signals and texture. And we'll get a lot of clinical experience, hopefully, in the near future, so we can train the network that can then be applied all over the place. Because it's trained to detect diffuse disease in the liver or in the kidney, or whatever.

TANAKA: As for clinicians, this multiparameter imaging is very interesting, so as early as possible we would like to have a clinical application possibility of this. After listening to your talk, there is one point that is not clear yet to me, so I would like to confirm that point. In your methods, the characteristics of the speckles will be expressed in the gray scales. Do you think that is the basic idea about your method?

THIJSSEN: I'm not sure if I followed all of your question. Your question that we want to express the parameters we estimate in gray scale? Is that right? Well, we're not sure. We still have the feeling that color would be helpful, that's first. And second is that we still need a good idea, or let's say a workable idea of how to combine parameters into one single image. We have thought of the way MRI works nowadays, that you have T1 weighted image and T2 weighted image and so on. We could do similar things, but the question is that in our experience, as I said before, one parameter would not be enough to highlight a tumor or an ischemic lesion in a heart. So we're aiming at combining all the information into one image, and that could be in gray scale, because some people like the gray scale. But on the other hand, it has been said before also now for the tissue Doppler imaging that was invented by Toshiba, I think–they do it in color, because it's a different parameter, so they don't confuse it with the original B-mode. So, my opinion is that color could be functional and practical, although I'm not one hundred percent sure about that.

TANAKA: I'd like to make another comment in Japanese. In whatever imaging, when you use waves such as ultrasound, depending upon the structure of the reflectors, interference might be there. So we can not avoid the speckles, I think. In your method, how are you trying to cope with speckles? What is your idea about hunting the speckles?

THIJSSEN: Well, there are two parts to my answer. The first is that part of the information we collect from the images is due to the speckling, and we call that texture analysis. I've shown that if, for instance, the number of scattering

6. Multiparameter TC and Image Processing

structures changes due to a disease, it could be an infarct, for instance, where there is a change in the structure of the heart muscle. If there are less scattering structures in that piece of tissue, then you find also a change in the speckle. So speckles can be information; they're not only disturbing. That's why we analyze them. If you talk about the attenuation and backscattering analysis, then you're right, the speckle is a kind of noise. And it's also present in the same way as it is in the image; it's present in the spectrum you have to analyze. So the only way to get rid of it is that you measure along a number of lines within the window. You make a spectrum for each line and then make an average of the spectra, so that you greatly remove noise from the speckle on the spectra. You never get rid of it completely, you're right, it's a disturbing factor, and it's inherent to ultrasound. So we cannot avoid it, I'm sorry.

HILL: Well, this is really a follow-up comment to Professor Tanaka's discussion of the presentation in color or gray scale. I believe it's true that a choice of displaying in color or gray scale will depend somewhat on the signal to noise ratio of the data that you're trying to present. Because the human eye and brain can to some extent integrate spatial differences in gray scale data. It may not be so good in integrating color differences over an area. So that certainly color has indicating flow and gray scale indicating anatomy. But to go to color because color sounds nice is not always the best thing for maximum perception. I'm speaking as an amateur, but this is my understanding of the perceptual science.

THIJSSEN: I'd like to have one final word about color. I think Kit Hill is completely right. The point is that we have a different brain functioning if an image is color, because we tend unconsciously to combine areas in an image which have the same color. This may be illustrated with what happened to the radiological images. Some companies offered the option to present CT in color, and then, very soon, it was discovered that this option was disturbing the perception of normal viewing i.e. of perceiving anatomical structures. So that's one big reason to be prudent about it. But as I said, you can also say that if you want to make a contrast between the normal and some extra information, you could do it better in color. But then it's an additional feature, and you have to restrict yourself to one or two or a maximum of three colors. That's what we discovered in our study as well. Too many colors is not helpful.

QUESTION: This is a Japanese question. Professor Dunn made a comment. In neural networks, I think the influence from the learning data is very large. After a large amount of learning data is accumulated, then the weighting function is going to be stable, and after stabilizing, this kind of weighting might have some physical influence. What do you think about this possibility of having physical influence of these stabilized informations?

THIJSSEN: I hope I understand the question. First of all, you're quite right that you need a lot of data to train a neural network. That's a complication. And then you need to do it many, many times. So let's say it can take a few days of computing time to train a network. Don't forget that. But then if it's trained well, you can use it. The point is that one of the reasons why we have started this study in Europe now that it's almost impossible to collect enough proven cases in one clinic; you need a lot, that's first. The second things is that we are kind of pulling ourselves by our own hairs because we generated data which have statistically the same characteristics as the data we have collected from patients. And we train the network with that. But it's not an optimal method; we realize that. The next point is that you suggested that after we've trained the network properly, you could look at the settings of the weights and try to get some information out of configuration that has resulted from that, and that's good idea. We have not done that until now. But I think it could be interesting to find out whether the weight factors have some practical use. That's right. Thank you.

Chapter 7

Elastography: A Method for Imaging the Elastic Properties of Tissue in vivo.

The medical motivations for looking into imaging of tissue elasticity are listed below:

- In many cases, a relationship exists between the presence of pathology and the palpable elastic properties of the tissue (e.g. breast, prostate, liver);
- Many breast and prostate cancers are routinely detected by palpation. Such cancers may or may not be seen by ultrasonic imaging;
- Detection by palpation is limited to relatively large, proximal hard nodules;
- Information contained in images of tissue elasticity cannot be obtained from any other diagnostic imaging modality.

In many cases a relationship exists between the presence of pathology and palpable elastic properties of tissues. For example, in breast cancer, a large percentage of breast cancers are detected by the patient. Many breast and prostate cancers are routinely detected by palpation. Such cancers may or may not be seen by ultrasonic imaging. It is another fact that many times you may not see such tumors with ultrasound or with mammography. Other times you can confuse what you see with benign lesions, for example, in the breast; some carcinomas look just like fibroadenomas. Detection by palpation, however, is limited to relatively large, proximal hard nodules. We have done many experiments where we

---

[*] J. Ophir, I. Cespedes, N. Maklad, and H. Ponnekanti

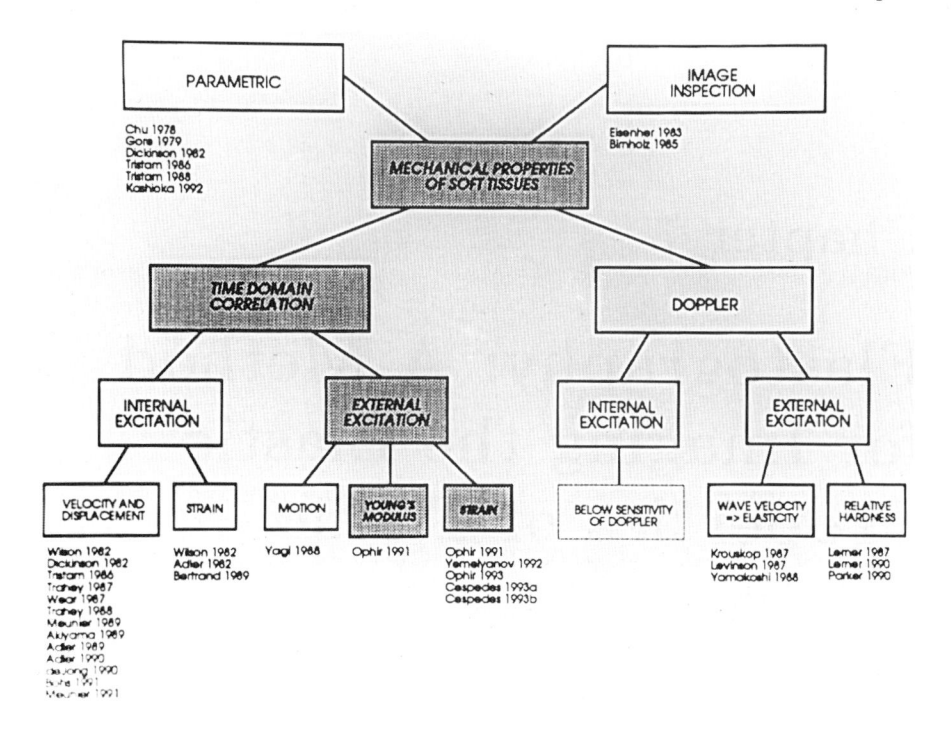

Fig. 7.1 Summary of the literature on the measurement of elasticity of soft tissues

have hidden a hard nodule in a gel phantom, and we tried to palpate it from the outside, and in most cases you can not feel anything, even though the nodule that you put in the phantom might be quite hard and large. Finally, information that is contained in imaging of tissue elasticity cannot be obtained from any other diagnostic imaging modality. Even though this parameter appears to be quite fundamental, you cannot get to it using any other sophisticated modality.

7.1 Previous Works on Tissue Elasticity

Figure 7.1 shows the kind of work that has been going on in the world in the last ten years in this area. This area basically has to do with mechanical properties of soft tissues. There have been basically four parts. One has been to simply inspect images undergoing motion, and then try to glean from that information about hardness and softness of material. There are some parametric methods that have also been looked at since about 1982. Even before that, in 1978, several groups have looked at different parameters of motion but did not compute elasticity directly.

There have been some Doppler techniques which have used either internal

7. Elastography: Imaging of the Elastic Properties

excitations or external excitation. In fact, internal excitations were deemed to be not strong enough to really get a good Doppler signal, but that could have been because of the wall filters that are in the machine that actually exclude this in favor of the motion from the blood. External excitations have been used by various authors here in the last five or seven years. Notably, the work at Rochester University by Lerner and Barker, where they have looked at relative hardness. They used a technique called sonoelasticity. There have also been some techniques in Japan–Yamakoshi and Sato, for example–and others which have attempted to measure the elastic wave velocity and relate that back to elasticity. Finally, we have a relatively large group of time domain correlation methods, which are subdivided into internal and external excitations. Using the internal excitation, investigators have looked at velocities and displacements of tissue as a function of the heart cycle, e.g. Dickinson and Hill, 1982, a fundamental paper. There were more papers from that group–e.g. Tristam, 1986 and 1988. In fact, 1982 was a productive year. Wilson and Robinson also have suggested that by using these time domain correlation methods, it might be possible to compute strain by looking at M-mode type information from internal excitation.

For time domain correlation due to external excitation, there have been some works going on–some in Japan by Yagi in Tokyo–including our group since 1990 attempting to actually look at strain directly, and Young's modulus estimation from that. O'Donnell's group at Michigan has reported similar work in 1992.

The domain of our work is covered basically by the shaded boxes(Fig. 7.1), i.e. we are doing time domain correlation, using external excitation, and attempting to measure strain and Young's moduli.

7.2 Theory

If we measure the strain along a single axis then the expression for the estimation of E(elastic modulus) is given as

$$E_i(x, y, z) = \frac{\sigma_{zz}}{\epsilon_{zz}} \left(1 - \nu \frac{\sigma_{xx} + \sigma_{yy}}{\sigma_{zz}}\right) \tag{7.1}$$

where ν is the Poisson's ratio, σ is the stress component on the element (x, y, z) whose subscripts indicate direction and ϵ_{zz} is the strain component in the z direction.

We can measure the strain along a single axis, and that is something which we are pretty much limited to, because the axis of the sound wave is where you have the maximum resolution. The other axes usually give very poor resolution, and when you attempt to incorporate off-axis information, you usually introduce more noise into the measurement than you can hope to gain. But in general, you can show from elastic theory based on some isotropic assumption that the estimation of the Young's modulus, E, of any point in space X, Y, Z (Fig.7.2), is possible. You have the transducer on top and you have the compressor on the

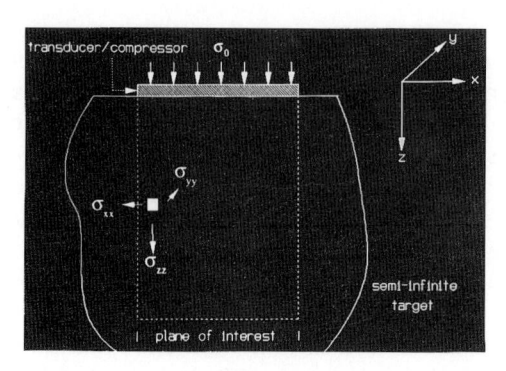

Fig. 7.2 The geometry of Young's modulus estimation of a point in an elastic solid, showing all the components of the stress tensor

bottom applying a downward stress. The modulus can be given by looking at the σ_{zz}, which is the stress downward, over the strain in the same direction. And then you have the second term in eq.7.1. If you forget for a minute about this term, then you can see that the Young's modulus is simply given as the stress over strain along the z axis. You can look at the second term as the error term, which involves the non-measurable stresses, and which subtracts from one. The Poisson's ratio is taken to be 0.5, or very close to it, for all soft tissues.

It is therefore possible to determine exactly the elastic modulus on a point-by-point basis of the object if you measure the strain along the axis of the beam. You measure the three stresses the σ_{xx}, σ_{yy}, and σ_{zz}, and you make an assumption, which is a very good one, about the value of the Poisson's ratio being close to 0.5.

With ultrasound, it looks like we can estimate fairly well the longitudinal components of the strain tensor, but you cannot measure stress. That is one of our fundamental problems. So if you want to be quantitative, you really can not do that very well. However, what we use is a solution that has been given for homogeneous materials under certain boundary conditions, where if you assume that the tissue is elastically homogeneous, and you know the boundary conditions, which we normally have a pretty good handle on, then you can use some solutions that have been in the literature for a long time. There is a solution due to Love(1929) which basically gives you values for all these three components of stress as a function of the distance from the top of the target in Fig.7.2. You can then plug in these value in the above equations.

Now admittedly, this is for a homogeneous material, but it turns out from some of our simulations that even if you use this assumption, you never encounter an error of more than a factor of two. On the scale of elastic modulus values that we think we will or are encountering in the body, a factor of two is rather small. We know from some preliminary work that the range of Young's moduli in the breast spans several orders of magnitude. So we feel that the error of a factor

7. Elastography: Imaging of the Elastic Properties

of two we incur an error by making some homogeneous assumptions about the stress field, is not unreasonably large. I will show later some artifacts that are encountered due to that assumption.

7.3 Elastography

The end product of elastography is a new kind of image which we call an elastogram, which is basically a two-dimensional image of the strain distribution in the object. In order to get a real quantitative image of the Young's modulus, you need to know the stresses and the strains. The strains are estimated here, where you basically look at a family of a lines, which are actually rf echo lines, before you compress the tissue. You apply rapidly a very small compression to the tissue, (on the order of 1% or .5%) and look at a second set of rf a-lines. We pair them up and do cross-correlation analysis in order to estimate the time shift of one relative to the other. And that, as I will show you a little later, leads you to get values or estimates of the local strain in the image. The term "local" means on the order of a millimeter or so in extent. If we assume that the stress is constant everywhere, then there is an inverse relationship between strain and Young's modulus. Thus if you can estimate or make an image of the strains, you have basically an inverse image of the Young's moduli. However because this stress does change, as I will show you later, more work must be done. In order to get a handle on that, we do measure the applied stress to the whole system. We can use some of the models in the literature, namely, those of Love(1929), to give you some theoretical stress distributions in homogeneous materials. From that, we can estimate the absolute axial stress, and that means all three components of stress, as you have seen from the previous formula. We then plug them into the equations that we have used before, and from that we compute the Young's moduli. The final product, as I say, is the elastogram.

Figure 7.3 is a finite element analysis picture of stresses that are incurred under a compressor. This is greatly exaggerated in order to see the detail, because usually we are compressing just 1% or so. The compressor is at the top, and you can see that under the compressor you typically get a kind of decay of the stress. There are very high levels of stress at the compressor's corners. You will see those later from some simulation results. But as you go further down, it is fairly well behaved. There is also some side-to-side variation, resulting in a "mechanical beam" of stress. Figure 7.4 shows some examples that we published recently in UMB, where you have the theoretical model from Love in 1929. Figure 7.4(a) shows the theoretical values, where you can see this kind of beam-like behavior of the stress. Figure 7.4(b) shows some experimental results that we produced at the lab that basically show you a very similar kind of a pattern, with an rms error of about 18% between the theory and the experiment(Figure 7.4(c)).

100   J. Ophir

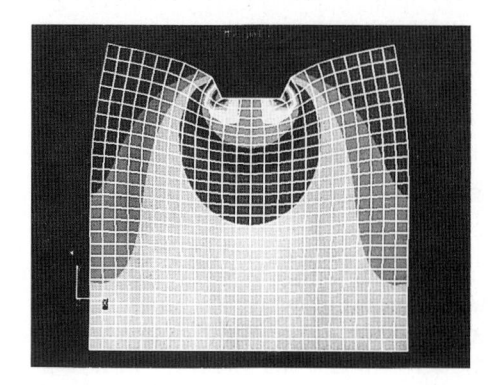

Fig. 7.3   A finite element simulation of the stresses under a flat compressor. Note stress decay with distance and high stress levels under the corners of the compressor

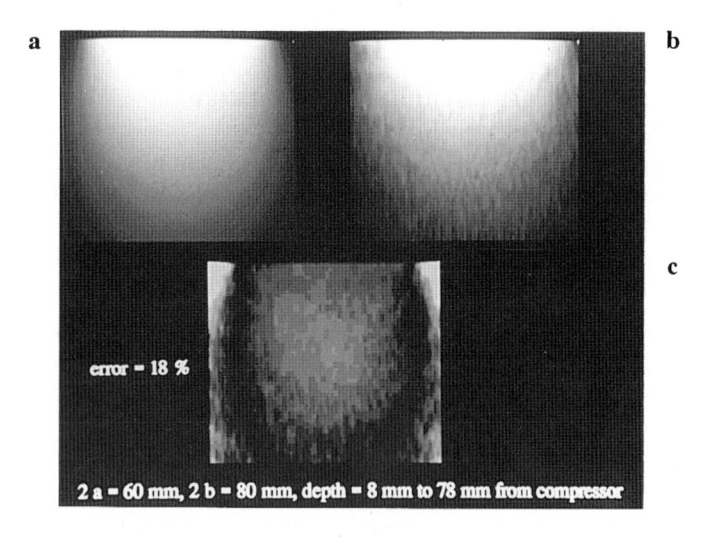

Fig. 7.4   The stress under a flat compressor (a)theoretical behavior using the formalism of Love(1929); (b)Experimental results; (c)RMS error between (a) and (b)

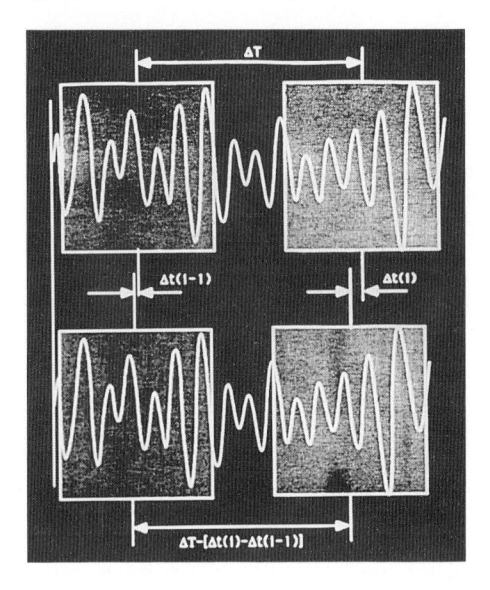

7. Elastography: Imaging of the Elastic Properties　　　　　101

Fig. 7.5 The estimation of strain. The time delay between congruent windows in two (pre- and post-compression RF signals) is estimated by cross-correlation. This is repeated for a second pair of windowed signals. The difference between the two delays divided by the original time interval between the windows is an estimate of the strain

7.4 Basic Technology

As mentioned earlier, the strain is estimated from the ultrasonic wave form, and the way that is done is shown in Fig.7.5. The figure shows a typical rf wave form, with two time gated windows. The centers of the windows are separated originally by some amount, ΔT. We then apply a compression to the whole system. What you typically observe is that there is some time shift, the sample will shift by a small amount, and that amount is measured by cross-correlation. The strain is actually estimated by looking at the difference between the new spacings and dividing this by the original spacing, and that will be the change in the size of the structure over the original structure, which if you assume a constant speed of sound, will then translate to the change in length of the structure over the original length, which is the longitudinal strength.

Figure 7.6 is typical cross-correlation function between the two signals. You can see that they are very close, and you are basically trying to estimate the time delay. What is also shown here is the sampling interval, Δt. Even though I have plotted this as a continuous function, all we really have to work with are the samples. And so you have a problem: if the peak is really between the samples, how well can you tell where that peak was, because all you have available to you are the samples? (See Fig.7.7.) We have looked at methods to estimate the peak

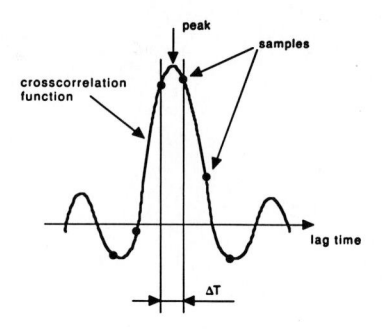

Fig. 7.6 A typical cross-correlation function between two ultrasonic RF signals

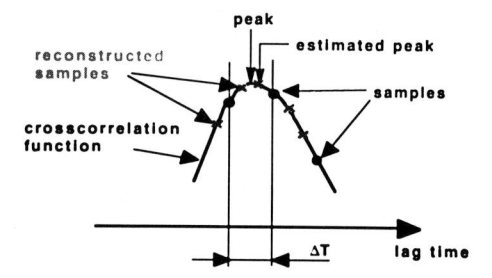

Fig. 7.7 Reconstructive interpolation method. Many new interpolated samples are generated in between the original samples by convolving the cross-correlation function with a sinc function

7. Elastography: Imaging of the Elastic Properties

of cross-correlation functions. There are basically two main techniques. One is to look at the actual digitized samples, and fit, for example, a parabola or a cosine or some such function to them. This works very well, it is very easy, but it produces large biases, so we have basically not used these methods much. What we use now is a reconstructive interpolation method(Fig.7.7), where we filter the spectrum by a boxcar-type of filter, which is the same as doing a convolution in the time domain with a sinc function, and what that does is to create many new samples in between the original samples (upsample); what we are doing is simply creating additional samples in between so we can approximate the peak location very accurately. The reason this is important because we are looking at very, very small time shifts. If we were compressing a structure which is 10 cm in size by 1%, then the whole structure goes down by 1 mm, which has to be divided over approximately 500 pixels. This means that 2 micron motions must be detected accurately.

The lowest possible errors that you incur in time-delay estimation for the location of the peak is given by the σ_{CR}, which is the Cramér-Rao lower bound for band pass signals. This is the best that you can do from a theoretical point of view. This lower bound is given in eq.(7.3), viz.

$$\sigma_{CR} = \frac{1}{2\pi f_o \sqrt{B\, T\, SNR}} \qquad (7.2)$$

$$\sigma_q = \frac{\Delta T}{\sqrt{12}} \qquad (7.3)$$

where the standard deviation of the error is inversely proportional to the center frequency f_o, and the square root of the signal to noise ratio SNR, and the time T bandwidth B product. So if you plug in some typical numbers, for example, 50 MHz sampling frequency, 20 ns sampling interval, center frequency 5 MHz, bandwidth of 3.75 MHz and signal to noise ratio of 48 dB, then you get the Chromér-Rao lower bound to be .024 ns, which is the theoretical lower bound on the estimate. If you digitize the signal then your digitization uncertainty is known to be equal to the ΔT to sampling interval over the square root of 12(eq.(7.3)). If you compute that for a 50 MHz digitization rate, you get a value of 5.7 ns. So you can see that the digitization noise, even at 50 MHz, will not let you attain the Cramér-Rao lower bound. This is again why you have to interpolate and be very careful how you do it, so that you can get close to the Cramér-Rao lower bound. Observe that interpolation is necessary even when the Nyquist criterion is met. Figure 7.8(a) shows the strain field under a flat compressor. Figure 7.8(b) shows the noisy elastogram; the noise is due to the uncertainty in strain estimation.

Figure 7.9 is the first elastogram we have produced. That noise is the result of your inability to exactly estimate time delay. When you put a tumor into an otherwise uniform area the effect can be explained on the basis of a one-dimensional model. If you look at the tissue as being composed of cylinders that

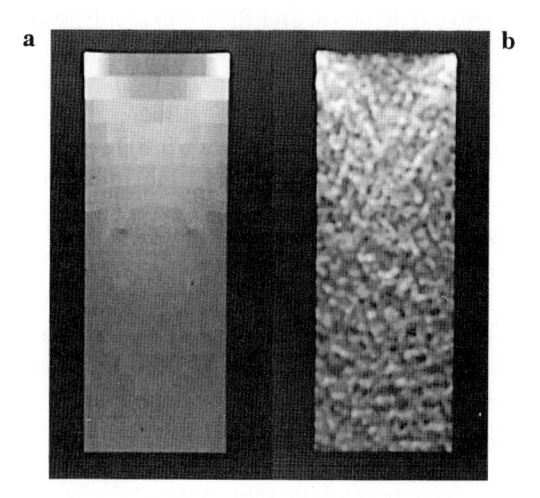

Fig. 7.8 Noise in the elastogram. (a) a simulated image showing the ideal strain under a flat compressor; (b) a noisy elastogram showing the noise added to the ideal strain distribution in (a) due to the uncertainty in time delay destination

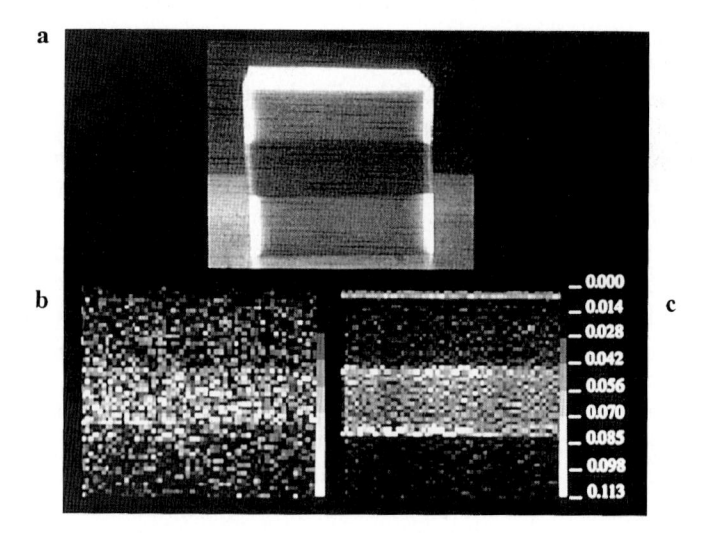

Fig. 7.9 The first elastogram produced in our laboratory(1990). The target is a composite sponge phantom. (a)photograph;(b)sonogram;(c)elastogram

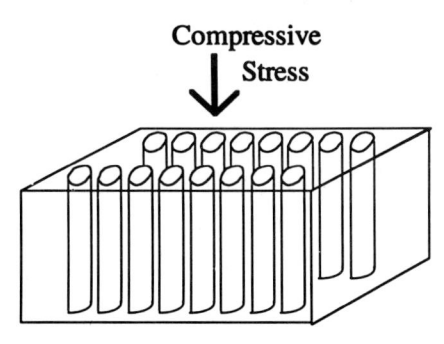

7. Elastography: Imaging of the Elastic Properties

TISSUE MODEL

Fig. 7.10 A simple model of tissue as composed of non-interacting, parallel elastic cylinders

do not interact with each other(Fig.7.10), then Fig. 7.11 will show you what happens here.

If the Poisson's ratio is zero, then you do not have any lateral interaction, and that is easy to explain then. You have, for example, three cascade springs of original length script l, and you have the nodes A, B, C, and D. When you compress the system by some amount $2\Delta l$, then if you do some simple statics calculations you will discover that each one of these springs has now compressed down to a new length of l, the original length, minus $2/3\Delta l$. If you sum up all the two-thirds Δl, you get back your global reduction in size, $2\Delta l$. And if you plot the strain, the strain would then be the two-thirds Δl, the change in the length over the original length. So it is two-thirds Δl over the l that you started out with. When you look at all those points now, A', B', C' and D', which correspond to depth, you basically get a constant strain. It simply tells you that all these springs are the same, and therefore they are all straining by the same amount (Fig.7.12).

In Fig. 7.13, we show a case where we substitute one of these springs by a totally undeformable spring. When we do the same experiment here, the very hard spring has an original length l, and it will simply be shifted down, but its length will not change. The first and the last springs, which are the softer springs, will now all compress by a full Δl, not two-thirds Δl. So now the combined deformation has to be picked up by two of them instead of three. If you plot the strain going from A' to B', C' and D', you get the picture in Fig. 7.14 where the strain is high in the soft springs and it drops to zero in this case where you have a very hard spring, or an undeformable spring, and then jumps back up in the bottom spring. But observe that now the strain in the soft springs, is higher than it was before. So the strain alone does not really characterize the spring.

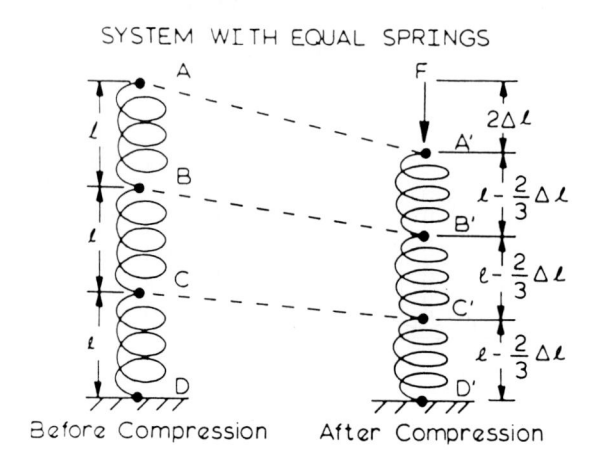

Fig. 7.11 A cascade one-dimensional spring model. All three springs have the same elastic modulus

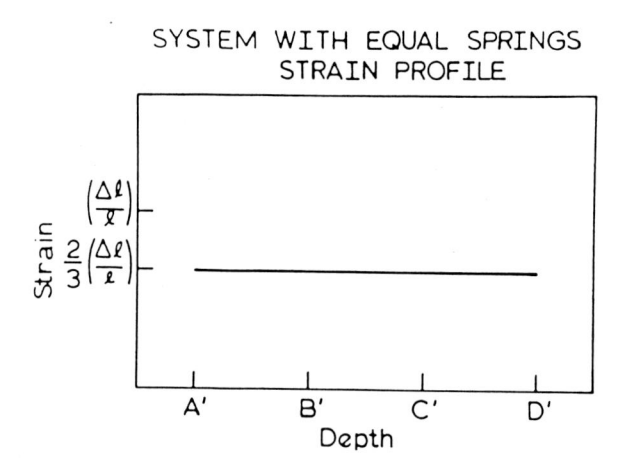

Fig. 7.12 The strain distribution in the springs of the previous figure. Note that the strain in all springs is equal

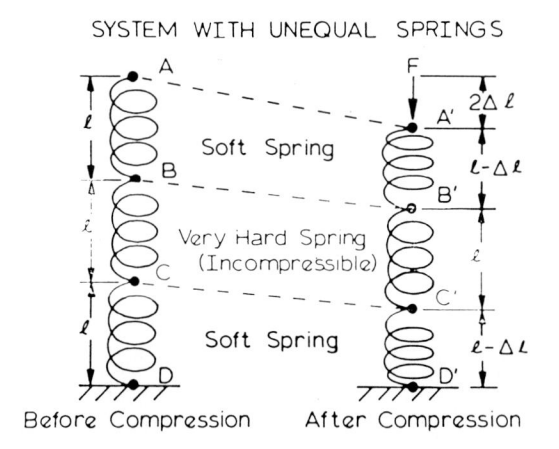

7. Elastography: Imaging of the Elastic Properties

Fig. 7.13 A cascaded one-dimensional spring model, where the central spring has been substituted by a totally rigid body

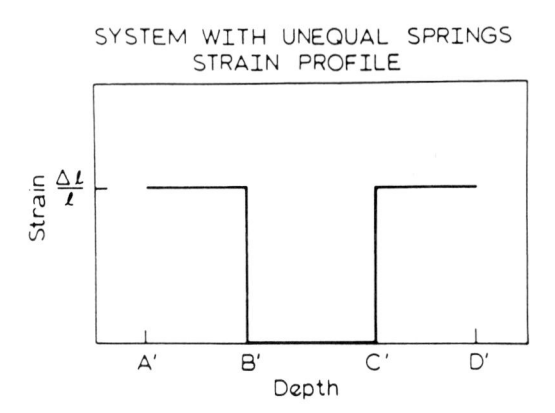

Fig. 7.14 The strain distribution in the springs of the previous figure. Note that the strain in the rigid spring equals zero

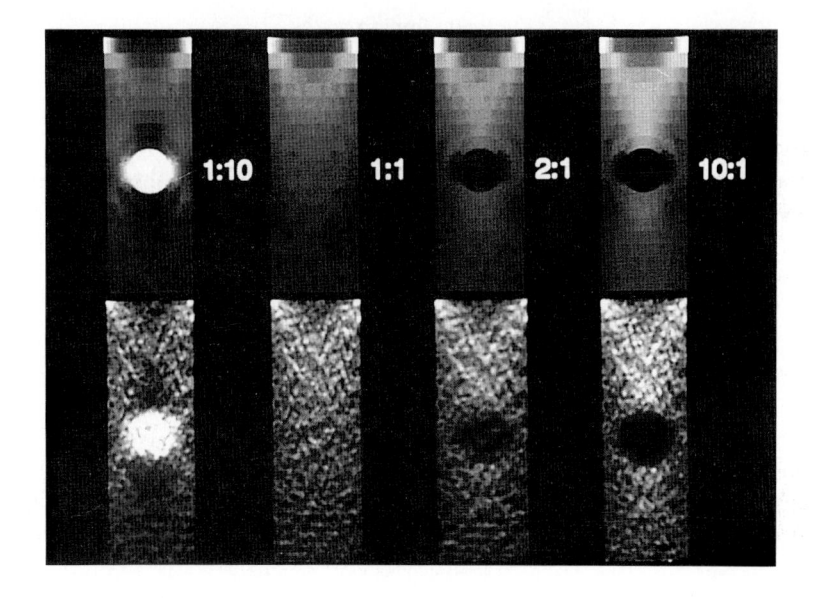

Fig. 7.15 Simulated strain images and elastograms of circular targets of various elasticity-constant levels

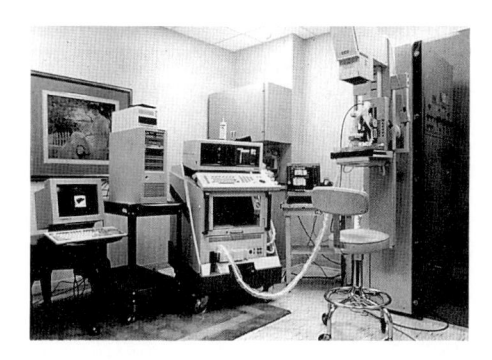

7. Elastography: Imaging of the Elastic Properties

Fig. 7.16 The clinical setup for elastography of the breast

Figure 7.15 shows images and elastograms in four elasticity contrast levels. These elastograms are simulated by putting many scatters in the locations of the finite element nodes and in between by interpolation, and then looking at what has happened to the rf as a result of compression, and producing elastograms from the rf signals. You can see that in both cases there are proximal and distal artifacts. In the images white is soft and black is hard. The presence of a soft structure produces hard front and back enhancements, and similarly, the hard structure produces proximal and distal elevated strain areas. The reason for that is that since you have a fixed deformation at the top, you may think of line-integrals going down vertically through the tissue here, and also through this "tumor". These line-integrals of the strains over the whole depth have to be equal to the total deformation. In other words, all line-integrals through all the strains in the vertical direction have to add up to the same displacement, because the integral of strain is simply the displacement.

We have constructed the system for the imaging of breast tissues, which involves a standard mammography machine onto which we have added a custom compressor (Fig.7.16). Figure 7.17 shows a custom mammography paddle. We have cut a hole in it, and we have put a standard 40 mm transducer array (a 5 MHz Diasonics transducer). It is driven by a stepping motor, which is computer controlled, and it can be swung out of the way so that you can take a mammogram first by putting the breast on the film cassette. You swing the transducer out of the way, get the mammogram, and then while the patient is still there and everything is stationary, you swing the transducer back, lock it into position over the breast, which now you can get to through an opening. By looking at the wet film and at these coordinates, you can look to see whether you have some suspicious area, and then you put the transistor over that and then you let this personal computer give the command to acquire the data from the Diasonics ultrasound machine, and then push the transistor down with this motor. Immediately afterwards, on the order of several tenths of a second, a second

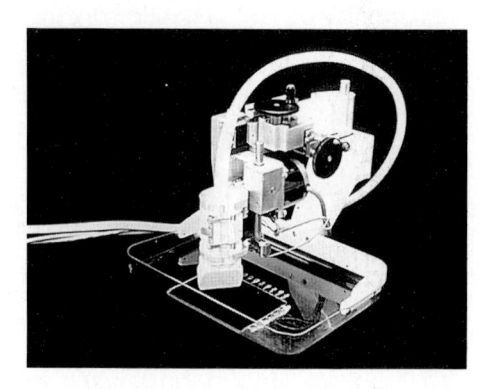

Fig. 7.17 A custom mammographic paddle used in elastography

ultrasound picture is taken. The acquisition is done digitally by a LeCroy 50 MHz digitizer, and the data are then transferred to the PC. Currently, it takes 2 or 3 minutes to process an image, but that can be done much faster using various techniques. So we think that it should be possible to arrive at real time or near real time performance.

7.5 Results

Figure 7.9 is the very first elastogram we produced in 1990. Panel(a) is the optical image of the target. It is a sandwich composed of two kinds of sponges that are submerged in water. The harder sponge is the white one, and the softer sponge is the black one. This was done before the system that I described earlier was in existence. The scans were performed with a single element 3.5 MHz transducer in a standard water tank, and it took over half an hour. It produces a relatively poor quality sonogram and an elastogram. But it is quite evident that you get two different kinds of pictures. It turns out fortuitously that the contrast in echogenicity was on the order of 6 dB, and the contrast in elasticity as measured separately in the physics lab was about 5 dB. So that the sonogram and the elastogram are quite comparable as far as contrast. You can see the typical ultrasonic speckle in the sonogram. It is a little bit blocky because of the low resolution. On the right you see an elastogram, and here again the black does not signify anything about the echo strength. It now means hard. And the white now means soft. One thing that you immediately see is that it appears that the signal-to-noise ratio is improved. Indeed the signal-to-noise ratio in the elastogram is on the order of 4, whereas in the sonogram it is on the order of 1.6 or 1.7. You can see a darkening in the far field in the elastogram. This darkening, or "hardening" of the far field should not happen. We also get

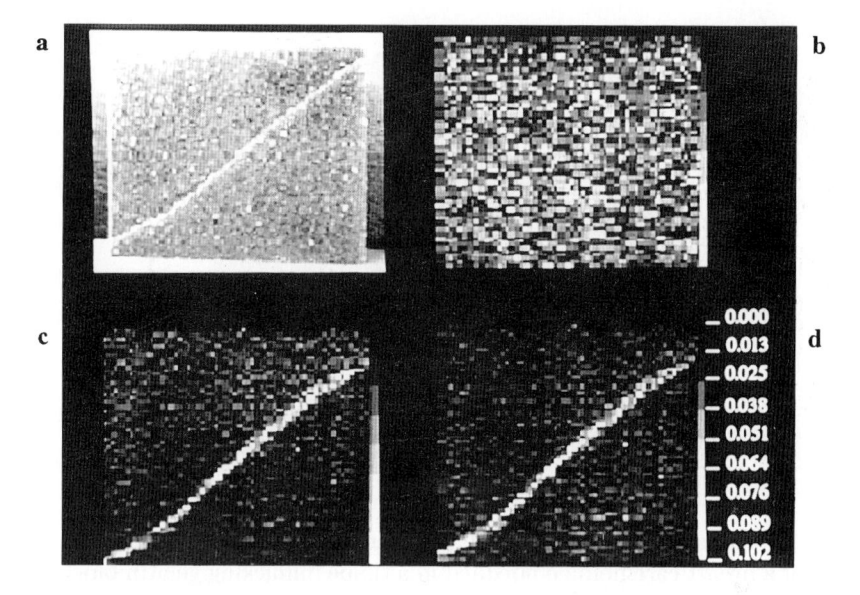

7. Elastography: Imaging of the Elastic Properties

Fig. 7.18 Diagonal cut sponge phantom (a)photograph; (b)sonogram; (c)elastogram; (d)calibrated elastogram with correction for the "target hardening" artifact

the white line between sponges, which is a sort of softening effect. We did not initially understand either of these effects. In order to try to explain some of these artifacts, we constructed another phantom, which consisted of a simple block of foam which we have cut in half diagonally with an electric knife, and we have then joined the two halves together(Fig.7.18). There was nothing here other than a cut. Now what happens at the cut region is that the intact pores of the sponge are severed and all you have left there are tentacles that are sitting up from the bottom and down from the top. Evidently, that localized area of the sponge becomes much softer than the rest of the sponge. So when you do a standard sonogram you see nothing; you simply get a speckle field. But when you take the same data and produce the elastogram, you see the presence of the cut very clearly. So apparently, when we are looking at the previous phantom structure, there are two types of foam which have been cut by necessity, and that area where the cut has been made weakens the foam at the edges and produces this relative softening of the material, as you can see very nicely. The other thing you can see is the artifactual darkening, which has to do with the fact that the stress does not remain uniform and in fact decays as you go away from the compressor. And so this stress decay or "target hardening", as we call it now, is basically an artifact which is similar in some sense to the attenuation-type appearance in sonograms. It is possible to correct for that fairly well, because we know what this function is from basic theory. A corrected one is shown in

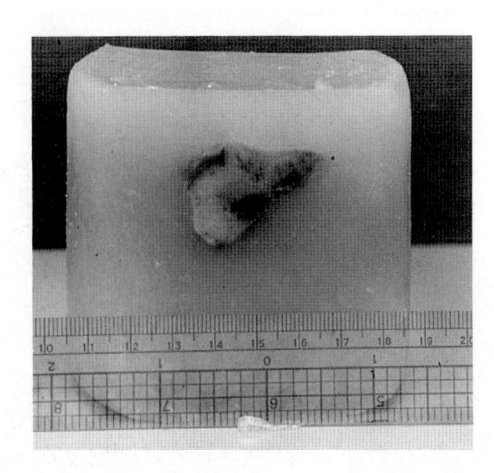

Fig. 7.19 Breast cancer in vivo. The photo shows a cut surface (in the scan plane) of a breast carcinoma embedded in a tissue mimicking gelatin block which contains acoustic scatters

Fig.7.17(d). The numbers on the sides are numbers that relate to the Young's modulus in the area, and they are given in inverse kilopascals. And the other thing that is informative about this particular picture is that you get some sense of the resolution of elastography as a function of depth. These pixels are 1 mm by 1 mm, so you see 1 or 2 mm type resolution in the focal area, and then you get some broadening in the far field and near field, similar to the "hourglass" appearance of focused ultrasonic beams.

Moving on from phantoms to in vitro tissues, Fig.7.19 shows a cut piece of breast cancer that we have embedded in a tissue mimicking gelatin block that contains scatters. We zoomed in on this area and produced a sonogram that you see on Fig.7.20 on the left, and the elastogram on the right taken from the same data exactly. We have added just enough scatters to this matrix of gelatin so that the contrast between the tumor and the gelatin was quite low, almost nonexistent. You do get a strong reflection from the top of the tumor, due to the normal incidence from the acoustic impedance mismatch between the gel and the tissue, which is an ultrasonic artifact. This case is an illustration of the fact that you can get little or no contrast echographically, but at the same time if you look at the elastogram, since this tissue turned out to be about five times harder than the surround, you get significant contrast.

Figure 7.21 shows our very first in vivo breast elastogram and sonogram pair. This is a cranio-caudal sonogram of a normal breast. The image size is about 40 by 45 millimeters. This is a typical normal breast of a 42 year old volunteer. The dark proximal area is the subcutaneous fat, which appears also in the distal area. There are also different kinds of white and black areas. It is never really clear, for example, if a particular black area is fat, just like the subcutaneous

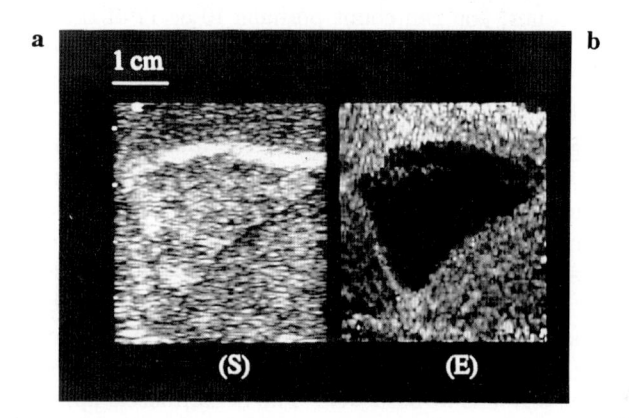

7. Elastography: Imaging of the Elastic Properties 113

Fig. 7.20 Images of breast carcinoma in vitro. (a)sonogram;(b)elastogram

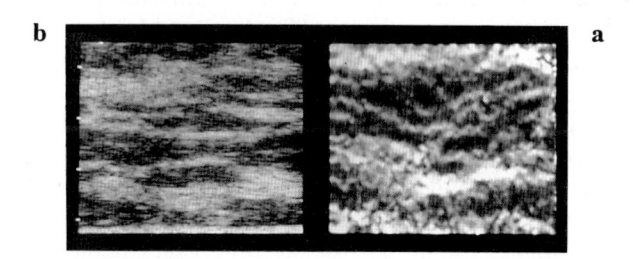

Fig. 7.21 Cranio-Caudal images of the normal breast of a 42 yr old volunteer. (a)elastogram;(b)sonogram

black area, or is it something else like a gland. If you look at the elastogram, one thing which is very apparent right away is that you get the subcutaneous fat, always the softest thing in the breast, showing up very well. The striking thing about this image is that it appears to have some regular structure of firm, rounded, or oval zones; you can count perhaps 10 or 12 of them. One does not see this kind of a structure in the sonogram, but perhaps there is only some suggestion of it. We think that these areas are consistent with the glandular structure of the breast, which is known to be firm.

The next image (Fig.7.22) is a picture of the lower leg (sonogram and elastogram). I will show you several more breast lesions later, and I wanted you to appreciate the fact that you can practice elastography also in other small parts, such as muscle. One can see both parts of the gastrocnemius. In the sonogram, one can see the soleus muscle coming down and bending around in 90 degrees, and the flexor halucis longus muscles is visible in the elastogram. And of course there are some fascia layers in between these muscles, which can be seen as well. In the elastogram, you get the gastrocnemius to show up very well. The soleus muscle can also be seen well. The flexor muscle appears quite harder than the soleus. You can see also see the fascia layers, which are supposed to be quite elastic. So there is good correspondence between the sonogram and the elastogram.

The next image goes back to the breast. We look at some pathology now Fig.7.23(c) is a mammogram of an 8 mm breast carcinoma. The sonogram and elastogram are in Fig.7.23 (a) and (b). This is an unusual carcinoma in that it is hyperechoic, and it is also greatly shadowing. In the sonogram, you lose some information distal to the lesion due to the attenuation. Just like in the normal breast, we see a laminar fatty layer on the top. The tumor is sitting near the fat. You can see the sharp borders of the lesion. This particular cancer, as most of them are, is harder than the fat.

Figure 7.24 is another breast cancer. This is a 3 cm cancer. Figure 7.24(c) is a mammogram of this cancer in a 62 year old patient and (a) and (b) are the sonogram and elastogram, respectively. The sonographic appearance shows a large hyperechoic area. Quite uncharacteristically, you have quite a bit of enhancement behind this particular cancer. Most cancers will actually be shadowing. That is one of the signs that people usually look at. So this was not very well understood, and when we did the elastogram we made some interesting observations. First of all, the only hard thing that we saw appears to be a rim that goes around the tumor. You do not see much of it on the back side, but proximally, you see a hard, thin region. The internal part of this structure appears to be soft. Indeed, this was then proven to be a mucinous degenerated large carcinoma, which means that the internal parts of this carcinoma have necrotized, and they were now fluid-like and rather soft. So this was quite different than the other cancer that we saw before, which was uniformly hard. You now have something that appears to have an envelope, which is perhaps an actively growing area, and then you have the central area here which is soft. Observe also what happens to the fat layer, to which we never pay attention. Remember

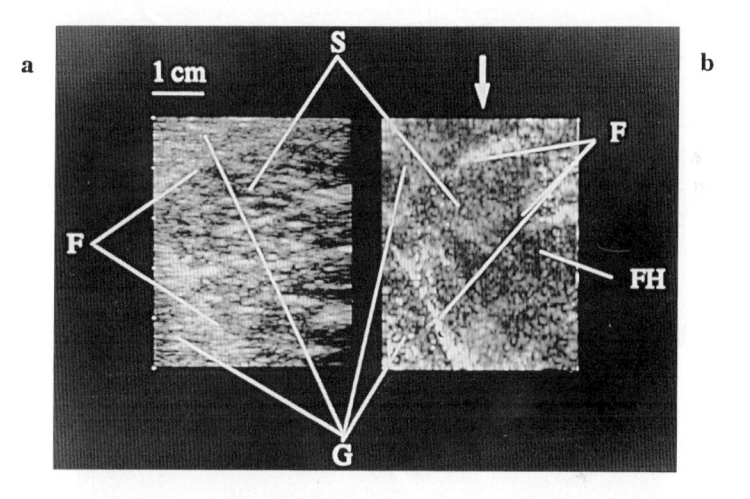

7. Elastography: Imaging of the Elastic Properties 115

Fig. 7.22 Images of the lower leg of a healthy male volunteer. (a)sonogram; (b)elastogram. G=gastrocnemius; F=fascia; FH=flexor halucis longus; S=soleus

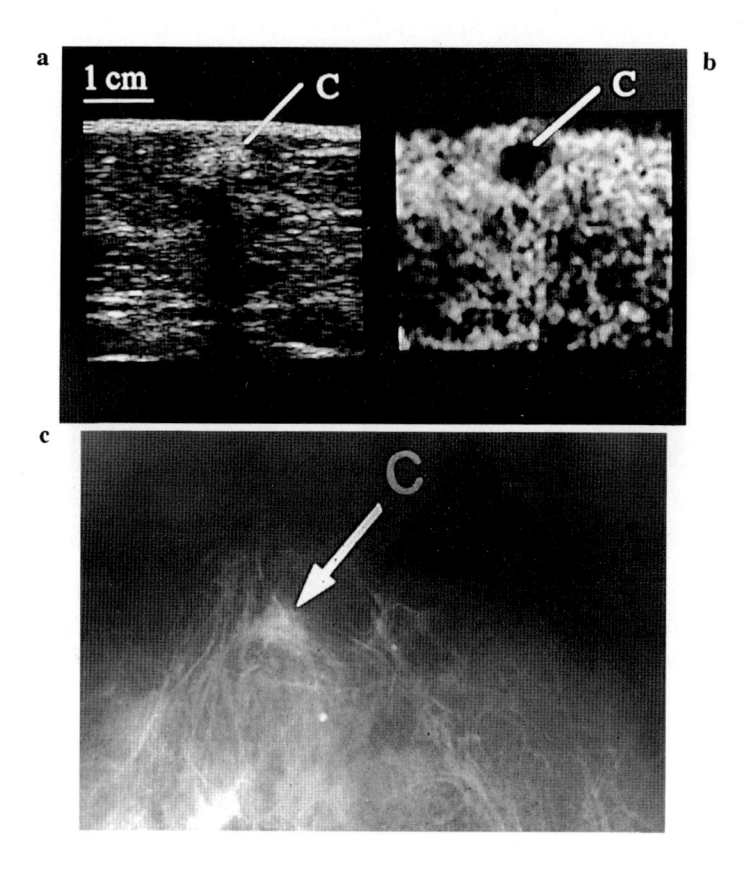

Fig. 7.23 An 8 mm breast cacinoma in vivo. (a)sonogram; (b)elastogram; (c)mammogram. Note the absence of shadows in the elastogram behind the hard, well defined lesion

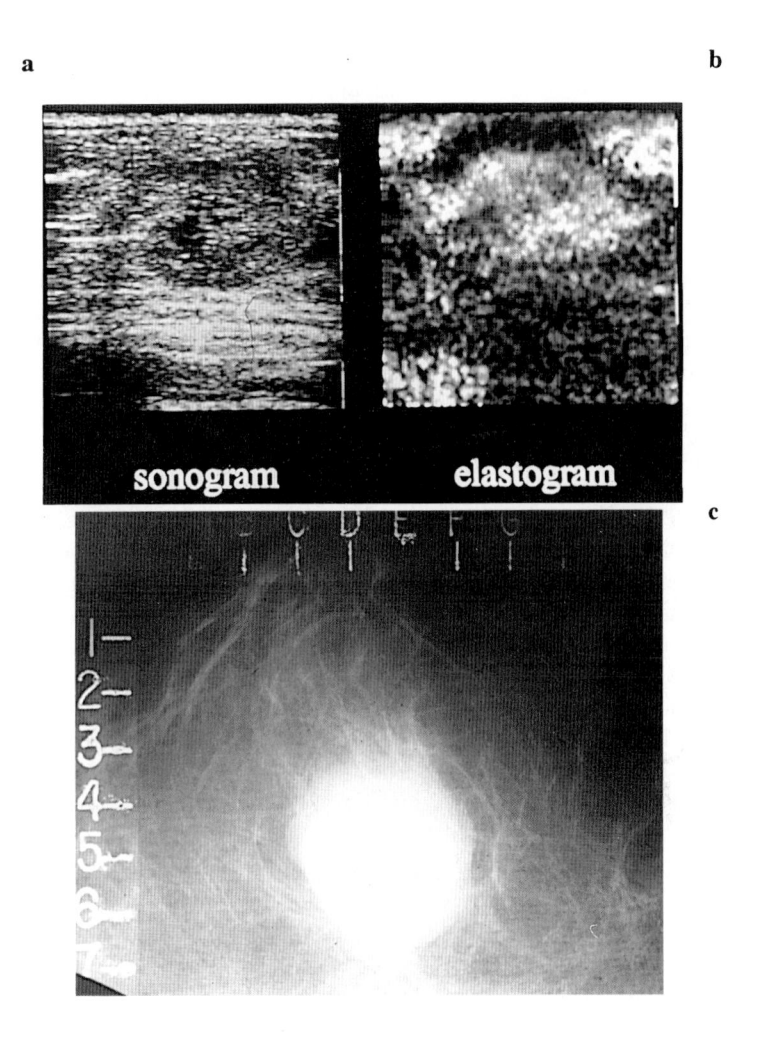

7. Elastography: Imaging of the Elastic Properties 117

Fig. 7.24 A 3 cm mucinous degenerated carcinoma of the breast in vivo. (a)sonogram; (b)elastogram; (c)mammogram. Note the destruction of the subcutaneous fat layer by the tumor, and the soft appearance of the tumor center in the elastogram

that in the previous cases the normal fat layer was laminar. In this case the fat layer, on what is left of it, is just two little pieces in proximal corners of the image. It looks like the tumor is invading it. By looking at what happens to the normal structure of the fat and in the presence of disease, we may get additional signs of malignancy.

The last case is a lesion that is quite irregular on the mammogram (Fig.7.25(c)). It has different parts to it. The sonogram (a) shows a hyperechoic area. You cannot see the fat here very well either. If you look at the elastogram (b), it is interesting because you see a larger structure of hardness, and we are not sure exactly what that means. It could be that the desmoplastic reaction around the tumor, even though it does not show up on the sonogram, may in fact harden a larger area than just the tumor itself, and that may show up here. There may be some degeneration inside the lesion. And then we see another hard region to the left. The interesting part is that the fat layer that we had talked about, instead of going across it now goes down between these two parts of the tumor, and then comes back up. This is an abnormal morphology of the fat layer.

7.6 Conclusion

Elastograms are influenced by many factors, many of which we still do not understand. It certainly is influenced by the acoustics, as you have seen. It is also greatly influenced by the signal properties and the signal processing that we use, and we are just now beginning to learn how to do accurate time delay estimations under different conditions. I am beginning to learn how to do accurate time delay estimations under different conditions. What I did not mention is that the signal does not only shift in time, but it also distorts. That is why we keep the compression low ($< 1\%$). When distortion occurs, the cross-correlation suffers greatly. So there are some methods and techniques to compensate for that, and those are actively being looked at. Similarly, the mechanics of the situation–the stresses and the strains–are also influencing the elastogram. And then there is some cross–talk among these parameters . So this is a very complex kind of a structure, and I believe we have enough work for many years to try to understand this, but I think you would agree that there is some potential here.

In conclusion, elastography is capable of imaging new tissue information in vitro and in vivo with reasonable sensitivity and resolution. It is possible to visualize elastic structures that are invisible on sonograms. Conversely there are some cases where you can see things on sonograms that you cannot see on elastograms. Potential applications would include the breast, muscle, prostate and perhaps other small parts.

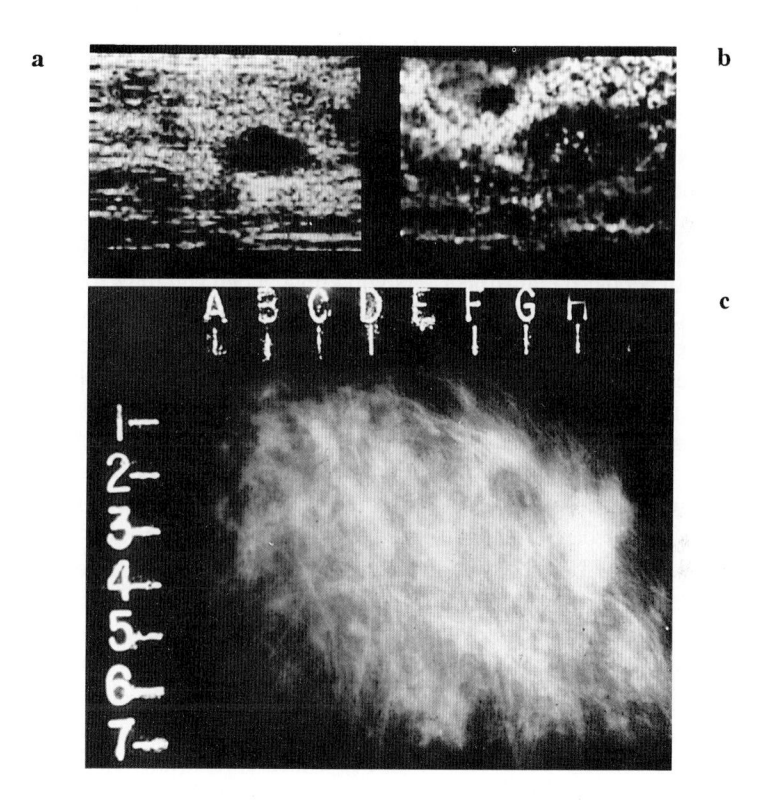

7. Elastography: Imaging of the Elastic Properties 119

Fig. 7.25 An irregularly shaped lesion in the breast. (a)sonogram; (b)elastogram; (c)mammogram. Note the appearance of one lesion in the sonogram and two lesions in the elastogram. Note also the distortion of the subcutaneous fat layer, which flows between the two lesions in the elastogram

Discussion

NITTA: Thank you, Dr. Ophir. This paper is open for discussion. We can have a couple of questions.

ENDO: I'm Endo from Kanagawa University. Thank you very much for your nice presentation. I will speak in Japanese. This method that you have explained–you estimate the stress at the beginning, I would say. In using this in vivo, are you estimating the stress by means of the finite element method?

OPHIR: Estimating the stress using the finite element method? The answer is yes and no. We can estimate the stress from the fundamental theory, but that is limited to some rather special cases. And yes, you could measure it from some of the finite element models, as well.

ENDO: So in the case of living organisms, you can calculate stress. And can I understand that you are calculating stress without the use of finite element method?

OPHIR: Do you calculate the stress also there from finite element methods? All the pictures that I've actually shown you have basically used very simple stress models which are derived from Love's work, which basically assumes that the tissue, as I explained initially, is uniform. This is sort of similar to how you characterize an acoustic transducer. As you know, when you have a transducer, it sonifies heterogeneous tissues. The sound field, as we have seen today, changes all over the place. But the manufacturer of the transducer always specifies in water or in homogeneous tissue, and then whatever errors are due to that we many times don't know about. We are doing something very similar here where we're characterizing our stress field either by finite element, by using the particular conditions of the situation, or from theory using the homogeneous assumption. As I mentioned, that causes errors or some artifacts, which in general are not bigger than a factor of two error in the resulting Young's modulus estimates.

THIJSSEN: First, I think that if you tried to estimate the time shift of two signals, you can only do that without a bias if the signals are identical. Otherwise, your correlation function will be biased. The second thing is that in one of your last slides you said that you confine or think you have to confine yourself to small organs. But I don't see any reason, a priori, why you shouldn't use the same method, for instance, for liver or any other kind of organ. Maybe it's a matter of applying a little bit more compression or whatever.

OPHIR: Yes, I agree that the time delay estimation that is usually described in the literature for using cross-correlation is just that. They're trying to look at two identical signals that are noisy and trying to estimate how far they have shifted from each other. And in our case, we do have distortions to the signal. Even if you only move by a very small amount, if it is speckled and not a deterministic target, the bandwidth will change and the spectrum will change, and the time domain appearance will change. And when you try to cross-correlate those two, you get a noisy estimate of time delay. So right now we are using small compressions, and the small compressions do several things. First of all, they make sure that that distortion is kept small, and therefore the bias in the

7. Elastography: Imaging of the Elastic Properties

cross-correlation function is small relative to the shift that you're trying to measure. It also assures, and that's something I didn't bring up in the paper, that there are no nonlinear effects of stress to strain relationships that we know very little about. But it is also known that under larger compression you might get changes in the Young's modulus due to that.

It's an interesting question about the size of the organs. There's no fundamental limitation here, except that what I didn't have to explain, that the decay of the stress under a compressor is relative to the size of the compressor. So if you have a very small compression–if you just push with your finger or with a small transducer–you will have a high stress field right underneath which will rapidly decay and will be gone. So in order to get the stress to go all the way down to deep structures, you have to obviously use a larger and larger compressor. You may reach a point where the compressor is so large that the footprint is too big on the body, and for practical considerations that may be bad. So the reason small parts are mentioned is that we can get away with standard, for example, 40 mm type transducers. And as you have seen from these images, you still have plenty of stress to work with. But it does decay, as you've seen in some of the other things. And it's really a matter of how small a stress can we measure over the noise to get to the depth we want. So it's a sensitivity issue. And you know how in sensitivity you can either up your TGC up to a point, or you can hit the transducer with more power. So in our case, you can either get better cross- correlation functions, or we can make a larger compressor, which is akin to increasing the power. Those two things will increase the penetration.

THIJSSEN: I'll make one more comment. As far as you've explained now, it's still more or less a matter of scaling, like you do with ultrasound in general. If you have a small organ, you use a small transducer at a high frequency. You talk about footprint size and things like that, but if you have a large organ, you work with a low frequency and you have a large transducer. So I think that if you talk about penetration of 5 cm for breast, you could talk about 15 cm in liver with a lower frequency. So there's no fundamental problem.

OPHIR: There's no fundamental problem, yes.

HILL: So far, your interest and other people's interests has been in applying this in tumors, but I wonder whether there are not other applications and if you've thought about them. Maybe you would have to use some of these applications using internal stimulus rather than external stimulus, but how about myocardial infarction? Is that a pathology that's likely to lead to changes in elastic properties of tissue? And would you be able to detect that, do you think?

OPHIR: I don't profess to be an expert on myocardium. I think Professor Miller has worked in this area for many years and in fact has told me and others too that he's been working on some stiffness matrices for muscles and so on. So perhaps you can bail me out.

MILLER: Well, there are many other experts in the audience who could say much more wisely than I could, but certainly to the touch, scar from mature infarct is very, very stiff compared to normal myocardium. And there are true experts here who could say more. So it's a very good candidate. And yes, as Jonathan alludes to, we have a very serious program to map out the elastic stiffness coefficients for heart under normal conditions and under a variety of pathologies. And I won't have time in my talk later on today to present this, but we are systematically measuring the five elastic stiffness coefficients required to characterize the uniaxial structure of a thin layer of heart, and from that we can generalize to thick layers which have rotations of those elementary layers. And we have already published C11 and C33. We're about to submit to Professor Dunn and Journal of the Acoustical Society of America C13. So I hope if someone in this room referees it, they'll give it favorable reviews. It's just about to be mailed. We'll do that when I get home. And C44 and C66 are very close. So that's where things stand.

TANAKA: It was indeed a very stimulating and interesting presentation, particularly for the cardiologists. Elasticity is related to arteriosclerosis and other disturbances, and therefore this is a very crucial issue. I have two questions. One is related to this elasticity. There is this structural dependent elasticity and material dependent elasticity. I think there are two types: elasticity of structure and of material. With what you have told us, how can you sort of discriminate between the two, or do you take the two as a whole? Could you enlighten us on this aspect?

OPHIR: The question from Dr. Tanaka states that there are at least two, probably two, kinds of elasticity. One is related to the material itself, and the other to the structure of the material. I'm glad you asked that question, because I think that in the morning, for example, some people described bulk modulus of elasticity and how this relates to the formulas that have to do with the speed of sound or ultrasound in the body. Maybe I can try to describe it in terms of a simple example. If you have a rod of steel, if you put an ultrasonic pulse on one end, it goes through and comes to the other side after several microseconds. The speed has to do with the fact that it is steel, pretty much. Now if you take that same rod of steel and you coil it up into a spring, now you can sit on it and bounce on it and it has a different elastic constant all together. You can still pretty much take the same ultrasonic pulse and propagate it through that coiled steel, and it will exhibit pretty much the same speed of ultrasound, but you will now have a kind of low-frequency structural component, or macro structural component, which has to do with the fact that this same steel bar which still has the same ultrasonic speed of sound has a new elastic modulus, which was acquired due to the shape and the form into which it was put. There is some literature out there that suggests that the type of elasticity that we're talking about here, which is a low- frequency type of elasticity really has to do with the macro structure of the tissue. So it doesn't have much to do with the molecules and the atoms and their elastic interactions, but rather the larger structure, which has to do with the tissue composition itself.

QUESTION: You have shown us a very beautiful picture of an elastogram of a tumor. For taking that picture, how long does it take for a particular patient, or tumor, say?

7. Elastography: Imaging of the Elastic Properties

OPHIR: I can tell you that right now, it takes a long time. Because the system that I showed is very cumbersome, it is basically a research tool. So it has not been optimized in any way at all. The fundamentals, though, are such that we should be able to move the transducer, take one image in about one- thirtieth of a second, move the transducer in another thirtieth of a second and take the second image in the same amount of time. So maybe in a tenth of a second, we should be able to get the two ultrasound images that we need. Right now, as I mentioned, it takes two or three minutes to process on a personal computer with an array processor, but we have already looked at some ways to compute one bit cross-correlation functions, which can be done much faster and basically still contain the phase information to the point where we can see very little difference between an eight-bit correlation function, or an eight-bit rf, and a one-bit rf for this purpose. So I think ultimately it should be possible to get a frame rate of maybe 10 per second or five per second or something like that.

Chapter 8

High-Resolution Measurement of Pulse Wave Velocity for Evaluating Local Elasticity of Arterial Wall in Early-Stage Arteriosclerosis

In order to diagnose cardiovascular system based on acoustic characteristics of the heart muscle or aortic wall, it is necessary to measure small vibrations on these walls [1]. However, the amplitude of the cardiac motion is very large during one beat period. Thus, it is very difficult for previously proposed methods or standard ultrasonic diagnosis systems to measure these small vibration signals. We have proposed, therefore, a new method to track accurately object motion, and by applying the method to the aortic wall or heart wall, small vibrations of these surfaces can be observed.

---

*Hiroshi Kanai, Ph.D.
†This work was supported by the Grant-in-Aid for Scientific Research (No. 06352032, 06304010, 05555108, 06213205, 06555113, 06750430) from the Ministry of Education, Science and Culture, Japan

8.1 A Method for Measuring Small Vibrations

In our method [2],[4], the demodulated received signal, which is reflected at the wall surface, is A/D converted at a high-sampling frequency, and in the resultant digital signal, the wall motion is tracked accurately. Figure 8.1 shows measurement results. The signal at the top of the figure shows the resultant movement, which almost coincides with the M-mode image superimposed in the same figure, even at low intensity. At the same time, from the measured movement of the wall, the small vibrations on the wall surface are obtained based on the Doppler effect, as shown in the bottom figure. By applying spectrum analysis to the resultant measured vibration, the heart muscle is characterized in each local area.

Figure 8.2 shows an example of an electrocardiogram, the heart sound of a normal young person, and small vibration signals detected by our method. These small vibrations are on the interventricular septum of the same normal young person. For the period of an interventricular ejection phase, the power spectra are obtained for three normal persons and for three patients with myocardial fibrosis by adriamycin injections. On comparison of these power spectra, there is a clear difference between them, at least in the frequency components less than 20 Hz. Moreover, for the patients with myocardial fibrosis by adriamycin injection, the power shifts to higher frequency, which corresponds to the change in the acoustic characteristic of the heart muscle.

The standard electrocardiogram, which was developed about 100 years ago, uses only low-frequency components, and they are not detectable continuously throughout the beat period. However, the small vibration signal measured by our method contains higher-frequency components up to 1 kHz, and contains local characteristics, and they are detectable in all stages of the cardiac cycle. Thus, by using these vibration signals, a new scientific field of noninvasive acoustic diagnosis of the heart and artery dysfunction may become available.

8.2 Application to the Diagnosis of Arteriosclerosis

In the diagnosis of arteriosclerosis based on pulse wave velocity (PWV) by the standard method, PWV is obtained from the difference in arrival time of the pulse wave propagating from the heart to the femoral artery using a microphone. Thus, the measurable points are limited to where the aorta exists near the skin surface and the distance between these points is several hundred millimeters. Thus, only the average PWV is obtained between these distant points. However, in the early stage of arteriosclerosis, fibrous areas several millimeters in diameter are scattered over the surface of the artery. The acoustic characteristics will change in the final stage of arteriosclerosis after growth of these fibrous spots. Thus, it is important, for diagnosis of the early stage of arteriosclerosis, to evaluate the local hardness of the artery wall. For this reason we applied the

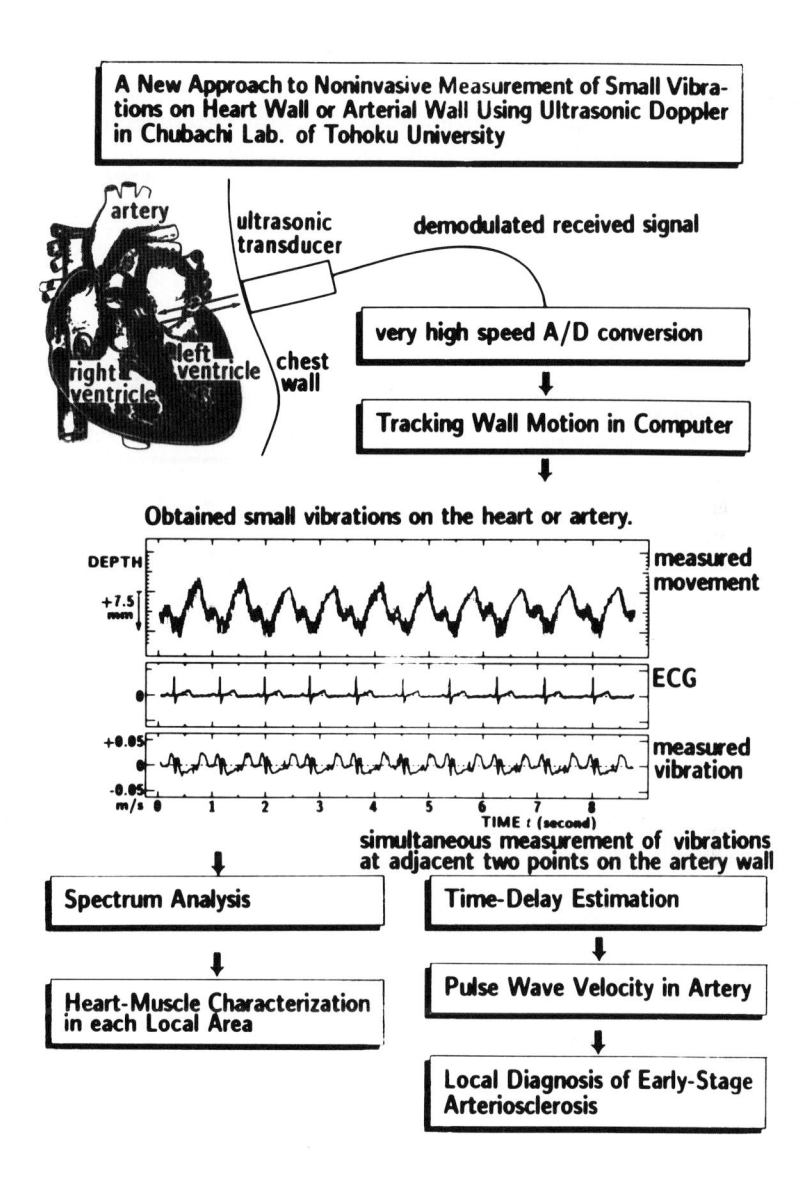

8. Pulse Wave Velocity for Elasticity of Arterial Wall

Fig. 8.1 The experimental results of the small vibrations on the aortic wall near the aortic valve of a normal young person

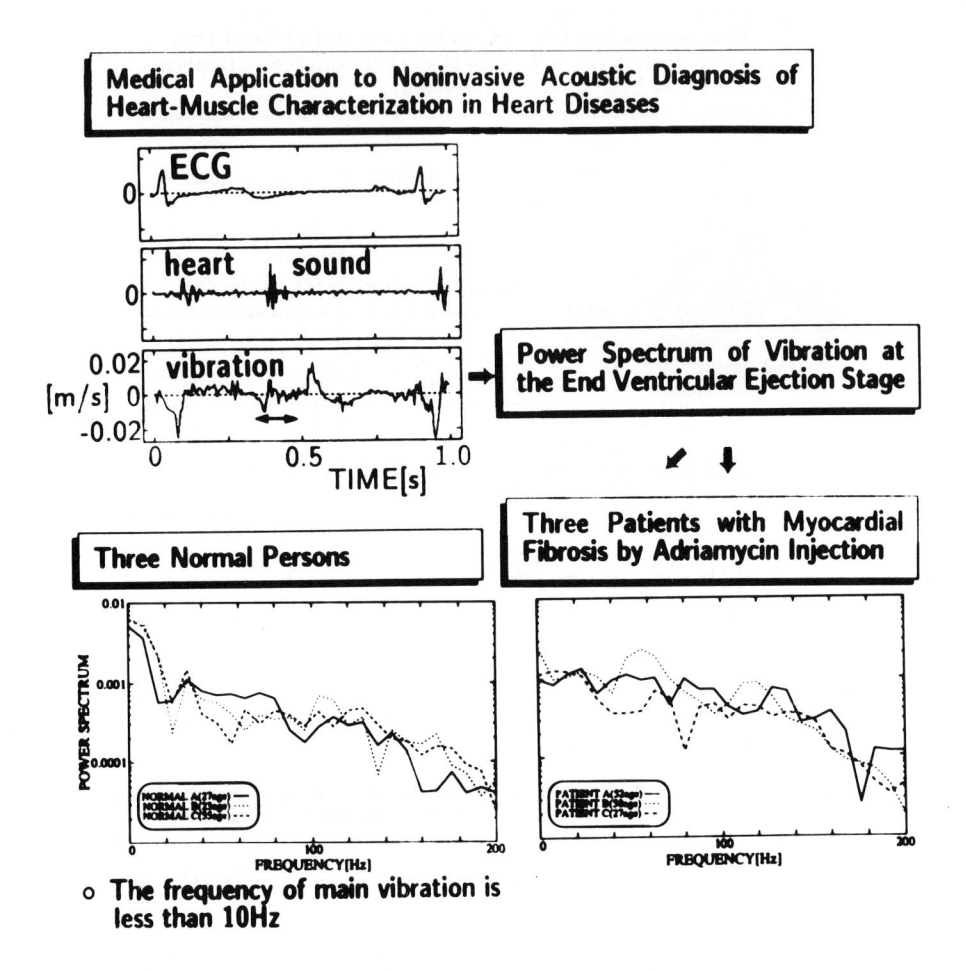

Fig. 8.2 Top: an electrocardiogram (ECG), the heart sound, and a small vibration signal on the interventricular septum of a normal young person, measured by our method. Bottom: The power spectra of three normal persons and three patients with myocardial fibrosis by adriamycin injection obtained by applying the spectrum analysis to the signal at the end ventricular ejection stage of the resultant vibrations

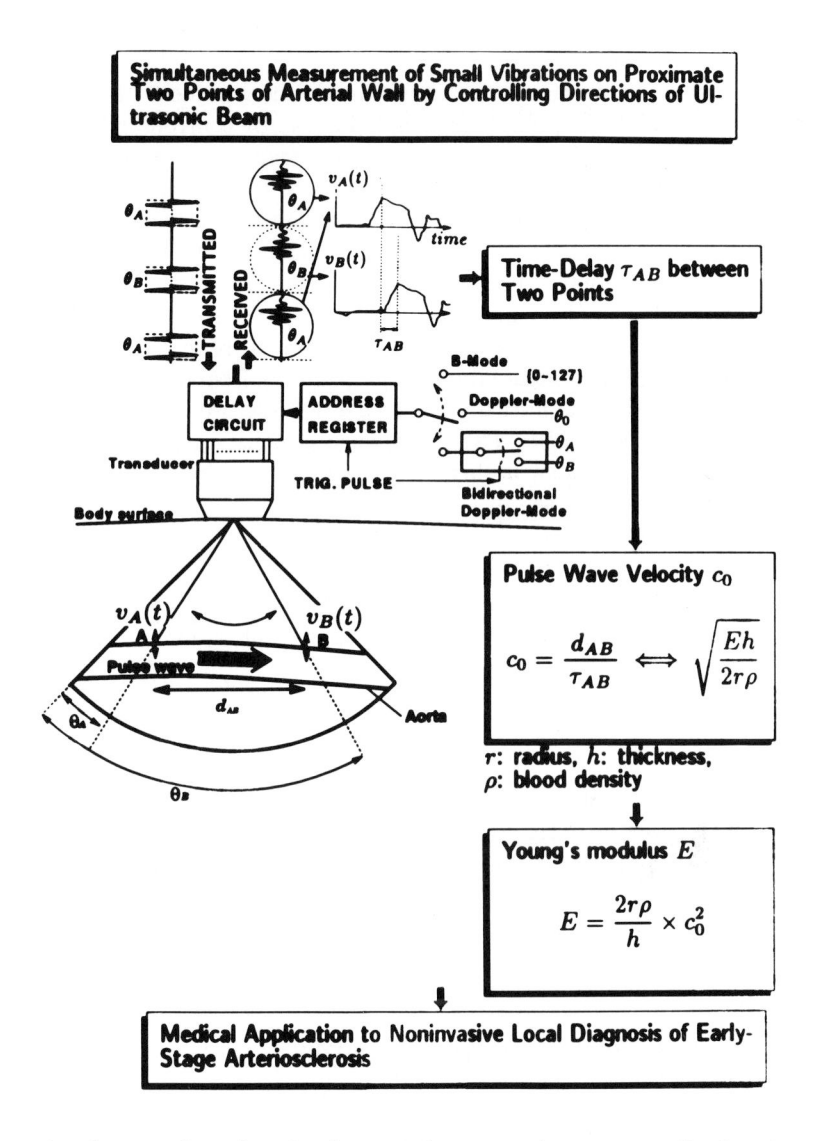

8. Pulse Wave Velocity for Elasticity of Arterial Wall

Fig. 8.3 A procedure for simultaneously measuring two small vibrations on the two points A and B of the arterial wall by controlling the direction of the ultrasonic beam. The elasticity of the wall is obtained by the time delay of the pulse wave propagation based on the Moens-Korteweg equation

proposed method to the simultaneous measurement of small vibrations of two adjacent points on the aortic wall [3]. From the transit delay time of the pulse wave of the small vibration between these points, values of the local elasticity could be obtained.

Figure 8.3 shows a procedure for simultaneous measurement of the small vibration on two proximate points of the arterial wall by controlling the direction of the ultrasonic beam. In the sector ultrasonic transducers employed, the direction of the ultrasonic beam is designated by the integer values. We added a small-sized circuit to the ultrasonic diagnosis equipment, by which we can control the direction of the ultrasonic beam, which is alternately transmitted in the two directions of the point A and point B on the surface of the aorta. From the reflected ultrasonic signal on point A and point B, the small vibration signals $v_A(t)$ and $v_B(t)$ are obtained by our method. From these vibration signals, the transit delay time τ of the small vibration propagating from point A to point B is obtained and, at the same time, the distance d of these two points is determined by using the M-mode image. The pulse wave velocity c_0, or the propagation speed of the small vibration, is then determined by dividing the distance d by the resultant time delay τ as

$$c_0 = \frac{d}{\tau}. \tag{8.1}$$

It is well known that there is a relationship between the pulse wave velocity c_0 and elastic Young's modulus E of the artery wall as given by

$$c = \sqrt{\frac{Eh}{2r\rho}}, \tag{8.2}$$

where ρ is the density of blood, and r and h are the inner radius and thickness of the artery, respectively. By measuring radius r and thickness h of the arterial wall, Young's modulus E of the arterial wall is obtained from the pulse wave velocity c_0 determined from Eq. (8.1). Since the Young's modulus E is evaluated for each local area of several millimeter range, it will be useful for noninvasive local diagnosis of the early-stage arteriosclerosis.

8.3 Water Tank Experiments

Before applying this method for *in vivo* measurements, we confirmed the accuracy of the proposed method by using a water tank as shown in Fig. 8.4. A silicone tube is excited by a small vibrator at the left-hand side in Fig. 8.4. At this time, the radial component of the small vibration propagating from the left-hand side to the right-hand side is measured at the two points, A and B, by the developed method using an ultrasonic transducer in the water. Let us denote the resultant vibration signals by $v_A(t)$ and $v_B(t)$, respectively. At the same time, the small vibration signals $v_C(t)$ and $v_D(t)$ are detected by the acceleration pickups at points C and D. The measured signal has dimensions of

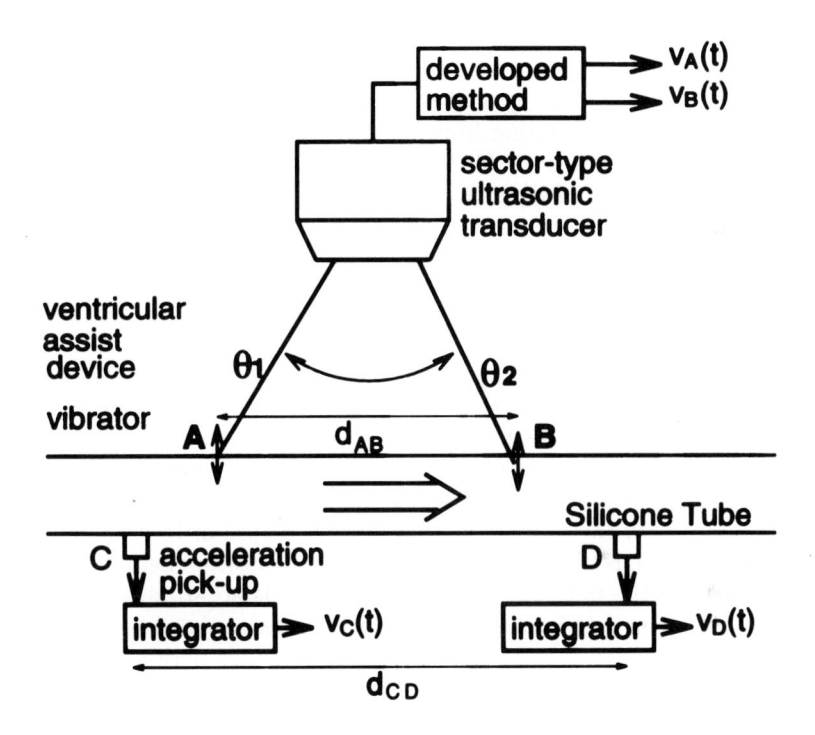

8. Pulse Wave Velocity for Elasticity of Arterial Wall   131

Fig. 8.4 The water tank experimental apparatus

acceleration. Thus, by integrating the output signal of the acceleration pickup, we obtain the velocity dimension signals $v_C(t)$ and $v_D(t)$, for which waveform can be easily compared with those of $v_A(t)$ and $v_B(t)$.

Figure 8.5 shows the experimental results using the water tank and the experimental apparatus in Fig. 8.4. The resultant output signals $v_A(t)$ and $v_B(t)$ measured by the ultrasonic-based system for the points A and B are shown in the top left figure. The waveforms $v_C(t)$ and $v_D(t)$, obtained by integrating the output signals of the acceleration pickups, are shown in the top right figure. These figures contain signals of about 20 periods. As shown in these figures, there is a time delay between the signals $v_A(t)$ and $v_B(t)$ or the signals $v_C(t)$ and $v_D(t)$. The bottom figures show the frequency characteristics of the transfer function $H_{AB}(f)$ and $H_{CD}(f)$ between these signals. From the gradient of the phase characteristics $\angle H_{AB}(f)$ and $\angle H_{CD}(f)$, each velocity, c_{AB} and c_{CD}, of the small vibrations propagating from point A to B, or point C to D is, respectively, obtained as 22.5 m/s and 22.2 m/s. Since the results nearly coincide with each other, the accuracy of the proposed method is confirmed adequately in the frequency range from d.c. to 500 Hz, and these values almost coincide with the values obtained in static experiments.

8.4 *In vitro* Experiments

Figure 8.6 shows the propagation velocity of the small vibration excited by the vibrator on the wall of a specimen of extracted human thoracic aorta [5]. The propagation velocity c_0 is measured for various values of diastolic pressure. For a normal young person, the velocity is almost 5 m/s for each section, as shown in the top figure. On the other hand, for an elder person of 78 years of age, the propagation velocity is about 7 m/s as shown in the bottom figure.

Moreover, from the pathological diagnosis, there is an arteriosclerosis lesion between points B and C in the same thoracic aorta of the elder person. For this lesion, the velocity is a little higher than for other normal part. Therefore, the propagation velocity of the small variations describes the local acoustic characteristics of the arterial wall.

The previously employed PWV depends on the diastolic pressure. In these experimental results, however, the small vibration excited by the vibrator is employed, and for this propagation velocity there is no dependence on diastolic pressure.

8.5 *In vivo* Experiments

Figure 8.7 shows an example of the *in vivo* measurement of small vibration on two proximate points, A and B, on the wall of the ascending aorta near the heart of a young normal person [6]. The signals $v_A(t)$ and $v_B(t)$, in the middle figure, show the measured small vibration signal on the points A and B. The distance between these points is 9.1 mm. The determined pulse wave velocity c_0

8. *Pulse Wave Velocity for Elasticity of Arterial Wall*   133

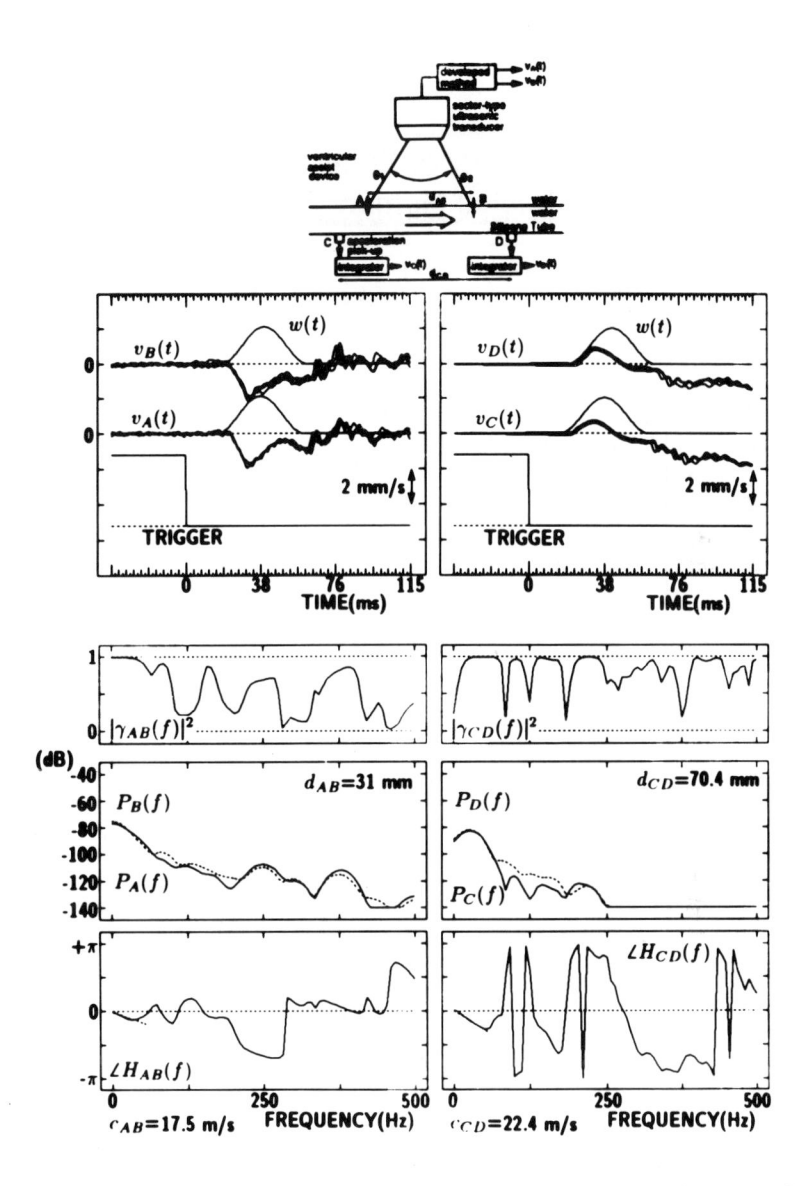

Fig. 8.5 The water tank experimental results. Top: the vibration signals measured at the two points, superimposed by referring the trigger timing of the driving pulse. Bottom: the transfer characteristics between the two vibration signals. Left: the results obtained by applying the proposed method to the output signal of the ultrasonic transducer. Right: the results using acceleration pickups

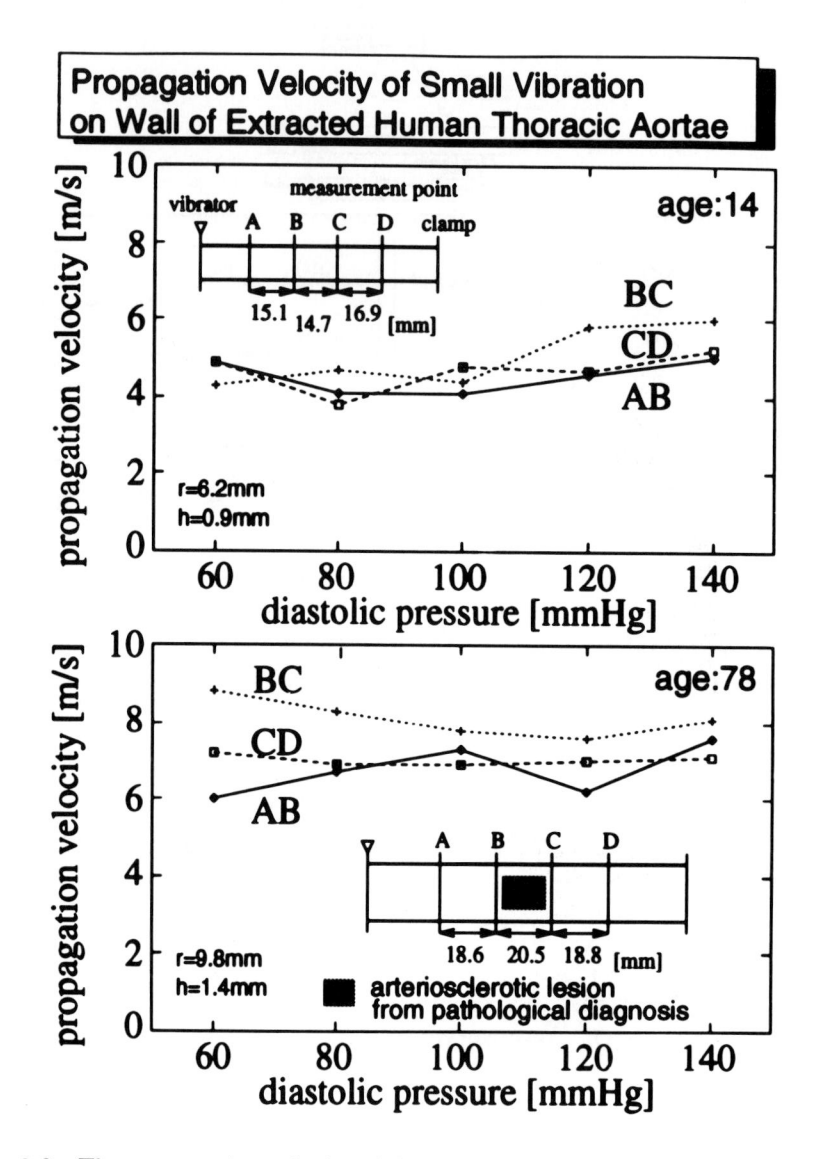

Fig. 8.6 The propagation velocity of the small vibration on the wall of extracted human thoracic aortae, for various values of diastolic pressure

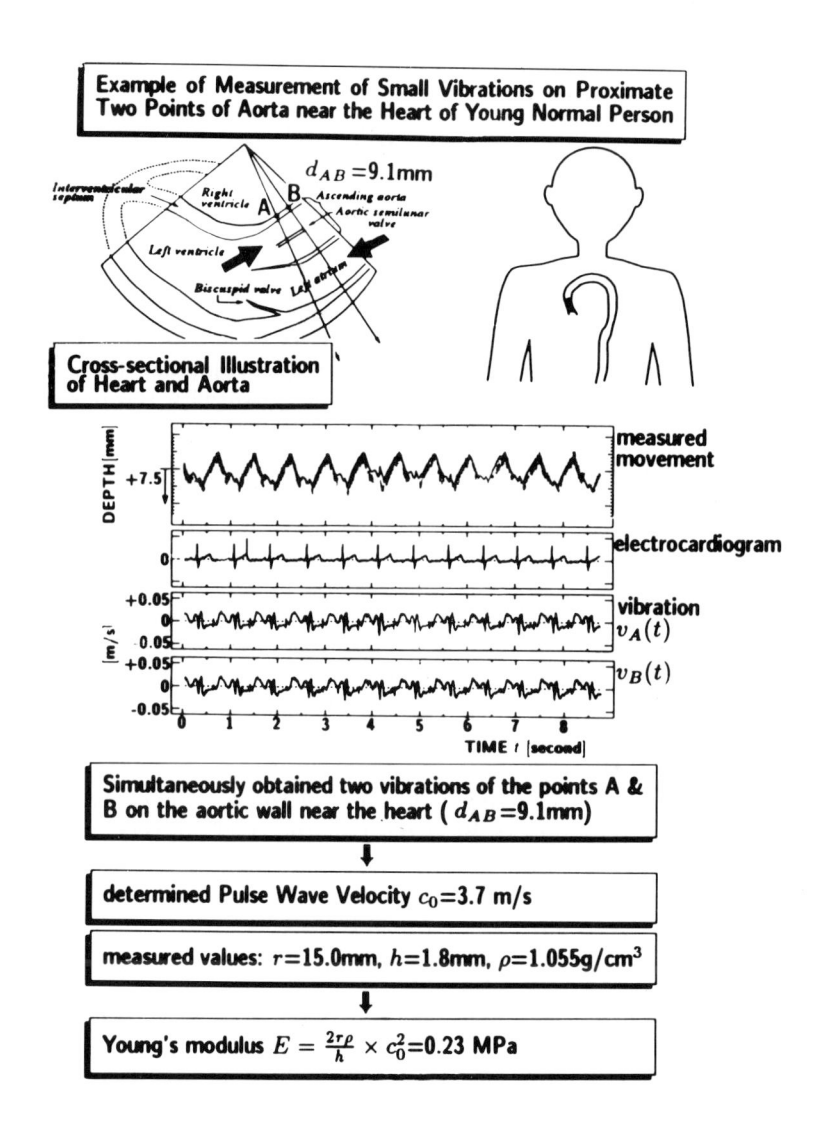

8. Pulse Wave Velocity for Elasticity of Arterial Wall 135

Fig. 8.7 An experimental result of the small vibrations on the aortic wall near the aortic valve of a normal young person

is 3.7 m/s, and by measuring the geometrical value of the artery wall (r=15.0 mm, h=1.8 mm, ρ=1.055 g/cm^3), Young's modulus is determined to be 0.23 MPa for this local area .

8.6 Conclusions

This paper proposes a new method for measuring small vibration signals at two adjacent points on the aortic wall. The transit time delay and the local propagation speed between these two signals are determined in the local area between the points. Since the small vibration measured by our method has higher-frequency components upto 1 kHz, the wavelength of the vibration is very short. Therefore, high spatial resolution of the diameter, which corresponds to the diameter of the fibrous area in early stage of arteriosclerosis is achieved. At the same time, from *in vitro* experiments of the extracted human thoracic aorta, local propagation velocity of the small vibration agrees with the pathological diagnosis. Thus, the obtained local acoustic property of the aorta is considered to be useful for noninvasive diagnosis of early-stage arteriosclerosis.

Acknowledgments

Prof. N. Chubachi, Mr. Hiroaki Sato, Mr. Ryoji Murata, Mr. Masahiko Takano of the Department of Electrical Engineering, Tohoku University, Prof. Dr. Yoshiro Koiwa of the First Internal Medicine, School of Medicine, Tohoku University, Prof. Dr. Fumiaki Tezuka and Dr. Mitsuhiro Takahashi of the Institute of Development, Aging and Cancer of Tohoku University are co-researchers of this work. The authors would like to thank Prof. Dr. Motonao Tanaka of the Institute of Development, Aging and Cancer of Tohoku University, and Prof. Dr. Floyd Dunn of Bioacoustics Research Laboratory of University of Illinois for their suggestions.

References

1. Kanai H, Chubachi N, Kido K, Koiwa Y, Takagi T, Kikuchi J, Takishima T (1992) A New Approach to Time Dependent AR Modeling of Signals and Its Application to Analysis of the Fourth Heart Sound. IEEE Transactions on Signal Processing 40:1198-1205

2. Kanai H, Satoh H, Hirose K, Chubachi N (1993) A New Method for Measuring Small Local Vibrations in the Heart Using Ultrasound. IEEE Transactions on Biomedical Engineering 40:1233-1242

3. Kanai H, Kawabe K, Takano M, Murata R, Chubachi N, Koiwa Y (1994) New Method for Evaluating Local Pulse Wave Velocity by Measuring Vibrations on Arterial Wall. IEE Electronics Letters 30:534-536

4. Kanai H, Satoh H, Chubachi N, Koiwa Y (1994) Noninvasive Measurement of Small Local Vibrations in the Heart or Aortic Wall. Proceedings of 16th Annual International Conference of the IEEE Engineering in Medicine and Biology Society Vol.16, 7.1-1

5. Takano M, Kanai H, Chubachi N, Koiwa Y, Tezuka F, Takahashi M (1994) A Study on Propagation Velocity of Small Vibrations on Aortic Wall. Proceedings of 16th Annual International Conference of the IEEE Engineering in Medicine and Biology Society Vol.16, 7.1-5

6. Murata R, Kanai H, Chubachi N, Koiwa Y (1994) Measurement of Local Pulse wave Velocity on Aorta for Noninvasive Diagnosis of Arteriosclerosis. Proceedings of 16th Annual International Conference of the IEEE Engineering in Medicine and Biology Society Vol.16, 7.1-6

Discussion

OPHIR (chairman): Since we ran late in the morning and we're running even later now, I was advised that what we should do now is have time only for one or two quick questions. And then since we are going to be visiting Professor Kanai's lab tomorrow morning, there will be more opportunity for questions at that time. So are there one or two quick questions right now?

THIJSSEN: I think this is a highly fascinating subject you have addressed here. You have given a formula to estimate the Young's modulus E from the wave velocity. Did you try to correlate your *in vitro* measurements with an assessment of the Young's modulus E obtained with another method?

KANAI: Yes, by the static experiments, we measured the stress–strain curve, and from the gradient, we measured directly the elasticity constant of the silicon tube. And as you said, there is a relationship between the pulse wave velocity and the elasticity constant. From the static experiment, we obtained the elastic constant and we derived the velocity. Then, we compared these velocities with the dynamic results, which almost coincided. Details of the experiments are described in Ref.[?].

THIJSSEN: So you could conclude that with this method you can have an *in vivo* assessment of the elastic modulus, in principle?

KANAI: Yes, that's right.

OPHIR: Thank you very much, Dr. Kanai. We will talk to you tomorrow.

Chapter 9

Some Relationships between Echocardiography, Quantitative Ultrasonic Imaging, and Myocardial Tissue Characterization

The goal of this work in tissue characterization is assessment of the heart based on analysis of sound return from the tissue, rather than on dimensions or motion. A number of approaches to the goal of tissue characterization have been proposed including qualitative texture analysis (such as "a sparkling appearance"), quantitative texture analysis (such as run-length statistics), and quantitative parameter-based imaging (such as imaging based on backscatter, attenuation, or velocity.) These approaches have been reviewed in detail in several articles [1-3]. In those reviews and in other reviews [4, 5] contributions from a number of Laboratories to the parameter-based imaging approach have also been summarized. Several additional recent references are provided here [6-13]. In this Chapter we will focus only on the work from our Laboratory in order to provide a coherent and systematic presentation of one approach to the goal of tissue characterization. The original manuscripts referenced provide acknowledgment of the creative contributions of many co-authors from our Laboratory, including especially Julio E. Perez, M.D. and Samuel A. Wickline, M.D. who have been essential collaborators in all of this work. Specifically, we will focus on an approach based primarily on real-time integrated backscatter imaging.

Integrated backscatter is a measure of the ultrasonic energy returned from a small region of tissue. Real-time two dimensional integrated backscatter im-

---
*J.G. Miller, Ph.D.

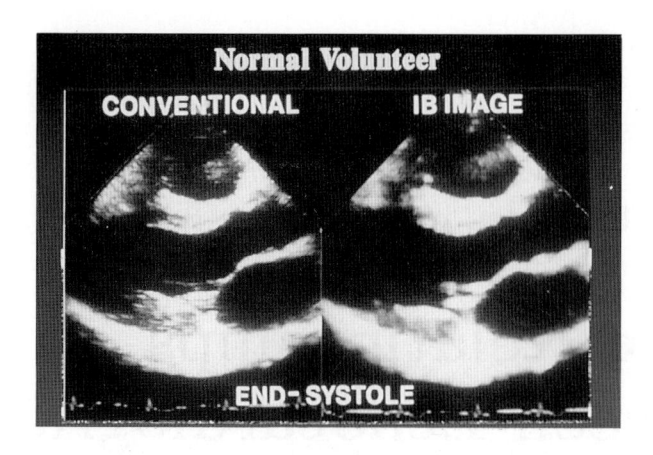

Fig. 9.1 Parasternal long axis views in a normal volunteer, illustrating the similarities between real-time integrated backscatter imaging and presently available echocardiography

ages are quite similar to those obtained in current echocardiographic studies, as illustrated in Fig. 9.1 [14-16].

We have identified three potential contributions that might result from investigations of myocardial tissue characterization: 1) Direct applications, specific pathologies that might be monitored as a step toward improved patient management, 2) Indirect applications, improvements and extensions of current echocardiography, and 3) Enhanced understanding of cardiovascular physiology.

A number of direct applications of tissue characterization have been proposed, including detection of altered myocardial structure and function due to acute ischemia and reperfusion, remodeling after infarction, diabetic cardiomyopathy, allograft rejection, dilated cardiomyopathy, hypertrophy and hypertrophic cardiomyopathy [1-3]. As described later in this Chapter, our Laboratory has recently addressed the issue of what sensitivity and specificity can be achieved in studies such as those. To appreciate the answer to that question, we first summarize properties of integrated backscatter observed in our Laboratory and confirmed by many independent studies as indicated in the references above.

In several standard echocardiographic views, myocardial integrated backscatter plotted as a function of time decreases during systole and returns during diastole [17-19]. (See Fig. 9.2) This is called the cyclic variation of myocardial backscatter. In acutely ischemic myocardium, the time or cardiac cycle averaged integrated backscatter gets larger ("brighter") by a few dB. In addition, the magnitude of cyclic variation is reduced from its normal value of perhaps 5 dB to a value that might be only about 1 dB. Furthermore, the lowest point on the integrated backscatter versus time curve, which for normal myocardial is

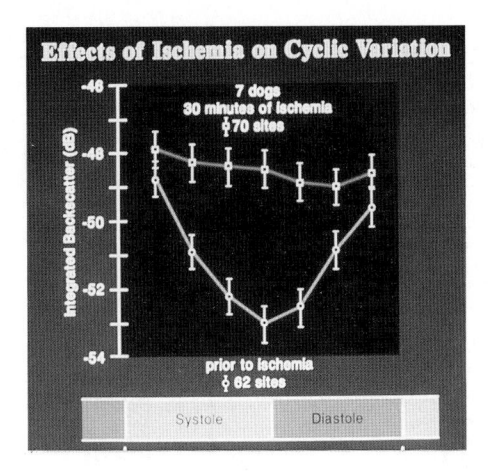

9. Relationships between Imaging and Myocardial TC

Fig. 9.2 Normal myocardium exhibits a cyclic variation of backscatter. Acute ischemia results in an increase in time averaged (cycle averaged) integrated backscatter, a reduction in the magnitude and an increase in the time delay of the cyclic variation[52]

usually near end-systole, is shifted later in time. We normalize this time delay relative to the systolic interval. This is reported as a normalized time delay of cyclic variation of backscatter and often has a value near unity for normal myocardium and somewhat larger values in specific pathologies [19, 20].

One parameter upon which tissue characterization might be based is the integrated backscatter time averaged over the heart cycle. This corresponds to simply the average brightness in a real-time two dimensional integrated backscatter image. This parameter typically goes up promptly by about 3 or 4dB with acute ischemia. It is substantially elevated (perhaps by about 10dB) in a zone of mature infarct [21, 11, 13]. (See Fig. 9.3, panel a.) A second parameter upon which tissue characterization might be based is the magnitude of cyclic variation, the bright to dark variation which is typically about 5dB in normal myocardium and less than that in pathological conditions. (See Fig. 9.3, panel b.) The normalized time delay of that cyclic variation of integrated backscatter, the time of the darkest point compared with the systolic interval, is a third parameter of interest [19, 24, 25]. As discussed below, we have explored the potential improvements in sensitivity and specificity that can be achieved by appropriate combinations of two (or more) such parameters.

Let us address the alterations in the cyclic variation of integrated backscatter associated with brief periods of ischemia followed by reperfusion using an index, the phase-weighted cyclic variation, which combines the effects of magnitude and time delay into a single parameter. Figure 9.4 illustrates the effect of the duration of prior ischemia on the recovery of cyclic variation after reperfusion

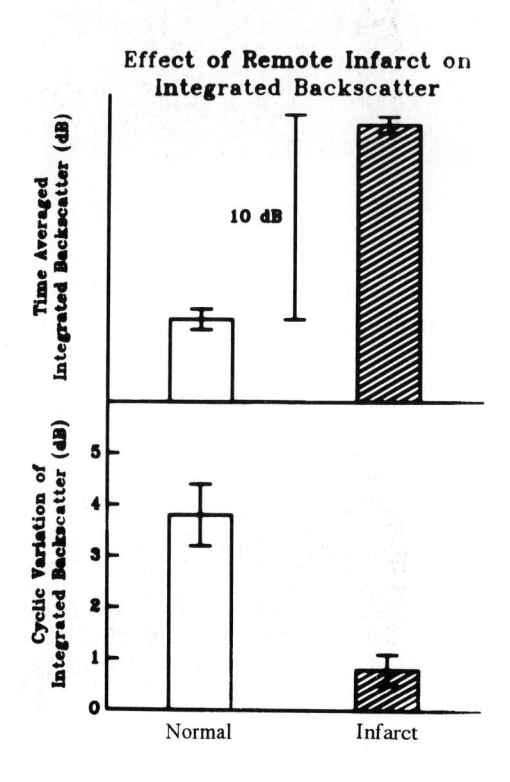

Effect of Remote Infarct on Integrated Backscatter

Fig. 9.3 Relative to apparently normal myocardium in the hearts of the same patients, zones of mature infarct exhibit substantially elevated time averaged (cycle averaged) backscatter and significantly reduced magnitude of cyclic variation of backscatter[53]

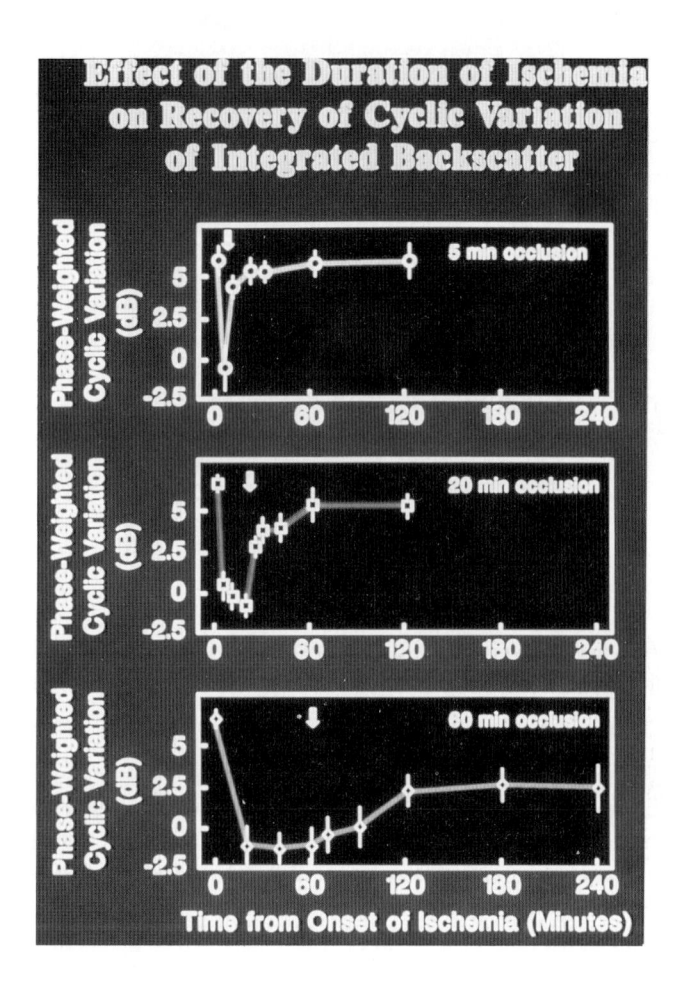

Fig. 9.4 Changes in an index that combines the magnitude and time delay of the cyclic variation of integrated backscatter for times after coronary occlusion and reperfusion. The onset of reperfusion is indicated by an arrow[26]

Recovery of Cyclic Variation and Wall Thickening after 15 Minutes of Ischemia
Canine Myocardium

(a) Magnitude of Cyclic Variation (dB): $y = (-1.8)e^{(-x/8.1)} + 3.5$

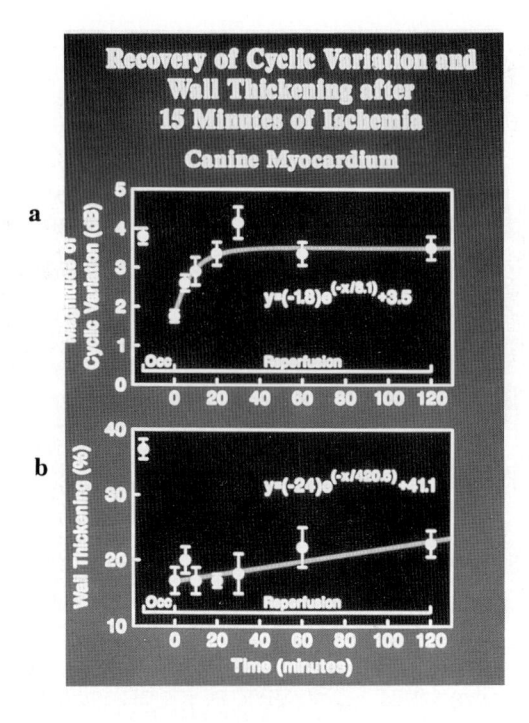

(b) Wall Thickening (%): $y = (-24)e^{(-x/420.5)} + 41.1$

Fig. 9.5 Recovery of the magnitude of cyclic variation (panel a) and the percent wall thickening (panel b) for 15 minutes of ischemia followed by reperfusion in dogs[27]

[26-28]. In panel a there was initially good cyclic variation which was reduced to a low value during occlusion, but returned promptly for only five minutes of occlusion. Panels b and c illustrate periods of occlusion of 20 minutes and 60 minutes, respectively. For 60 minutes of occlusion followed by reperfusion, there is only a very gradual recovery, presumably reflecting the longer period of ischemic insult.

One potential role for tissue characterization is in monitoring the effects of reperfusion induced by either chemical or mechanical means in patients with acute infarction. Figure 9.5 illustrates a comparison of the rate of recovery of an index based on the cyclic variation of backscatter with regional wall thickening in an experimental animal preparation [27]. The recovery of the magnitude of cyclic variation (panel a) after 15 minutes of ischemia is relatively prompt after reperfusion. In contrast, wall thickening (panel b) recovers much more slowly, presumably as a result of stunning of that myocardium. Thus the return of cyclic variation might be evidence that reperfusion was successful, available to the clinician prior to any evidence of significant recovery of regional wall thick-

9. Relationships between Imaging and Myocardial TC 145

Ultrasonic Tissue Characterization In The Diagnosis Of Ischemic Heart Disease

Normal Myocardium
- Normal Cyclic Variation
- Normal Absolute Backscatter
- Normal Frequency Dependence

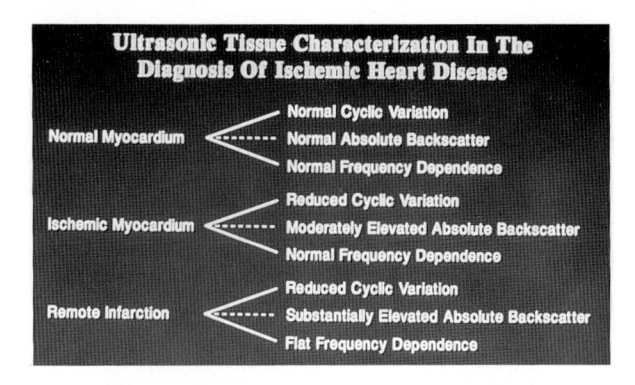

Ischemic Myocardium
- Reduced Cyclic Variation
- Moderately Elevated Absolute Backscatter
- Normal Frequency Dependence

Remote Infarction
- Reduced Cyclic Variation
- Substantially Elevated Absolute Backscatter
- Flat Frequency Dependence

Fig. 9.6 The potential role for tissue characterization in acute and remote ischemic injury

ening. Evidence of similar behavior in patients undergoing reperfusion provides encouraging support for this hypothesis [28].

Figure 9.6 summarizes a hypothetical approach to the role of tissue characterization in the diagnosis of ischemic heart disease. Failure of a localized region of myocardium to thicken might be consistent with acute ischemia or alternatively with old infarct. Acutely ischemia tissue could potentially benefit from reperfusion, in contrast with mature infarct. The frequency dependence of backscatter provides another index that may aid in the differentiation of old scar and acute ischemic injury, as illustrated in Fig. 9.6.

Figure 9.7 summarizes experimental studies carried out to illustrate the relationship between the frequency dependence of scattering and the size of the dominant scatters [29]. From theoretical considerations, the rate of rise with frequency of the backscatter coefficient is expected to exhibit a power law dependence with coefficient n, where n is approximately 4 for scatters very small compared to the ultrasonic wavelength and where n decreases to progressively lower values as the size of the dominant scatter increases. This relationship is shown explicitly in the panel a of Fig. 9.8, where the backscatter coefficients for the 200 micron and 7 microns phantoms of Fig. 9.7 are presented. The rate of rise is notably reduced for larger scatters. Panel b of Fig. 9.8 depicts the same quantities for normal myocardium and for 6 week old infarct [23]. The slower rise with frequency (as well as the increased overall magnitude) in zones of infarct presumably reflect the larger average scatter size in the collagen-rich zone of infarct. A comparison of Figures 9.6, 9.7, and 9.8 summarizes how tissue characterization might be useful in the diagnosis of ischemic heart disease permitting the segregation of normal myocardium, zones of acute ischemia that might benefit from reperfusion, and zones of mature infarct which presumably would not.

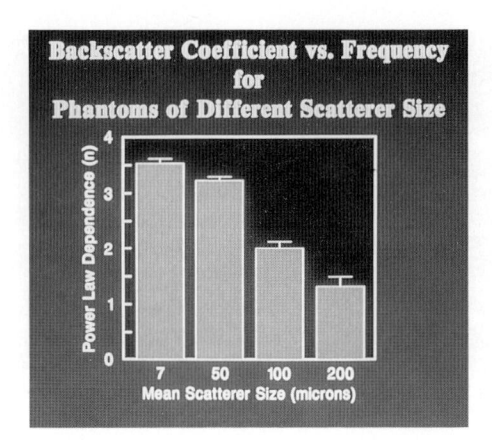

Fig. 9.7 Measured frequency dependence (power law exponent n) for four tissue mimicking phantoms of varying sizes[29]

Although the examples illustrated above suggest that there may be a role for tissue characterization in diagnostic cardiology, definitive conclusions regarding the utility of such an approach require careful analyses of questions such as "What kind of sensitivity and specificity can be achieved with tools of this sort?" (See Fig. 9.9) More appropriately, one should ask questions regarding the performance of tissue characterization in the context of Receiver Operating Characteristic (ROC) curves [30]. In evaluating a diagnostic test, one can trade improved sensitivity for decreased specificity, and the reverse. (See Fig. 9.10.) Such tradeoffs result from choosing to operate at a different point on the receiver operating characteristic, or ROC curve. The area under the ROC curve is a useful measure of quality, with larger areas indicating better diagnostic tests.

An ideal diagnostic test would have an area under the ROC curve of unity. At the opposite extreme, the corresponding area obtained by flipping a coin, a totally random outcome, would have an area of 0.5. Some authors have advocated reference to a widely accepted standard of diagnostic performance, that of the pap smear, which under optimal conditions is believed to exhibit an area under the ROC curve of 0.87. So our goal is to see what tissue characterization can do in comparison to that standard. (See Fig. 9.11).

9.1 Tissue Characterization in Insulin-dependent Diabetics

To examine the outcome of ROC analysis on myocardial tissue characterization, we selected data from a previously published study of the hearts of diabetic patients [31-33]. These published data were then reanalyzed to examine the re-

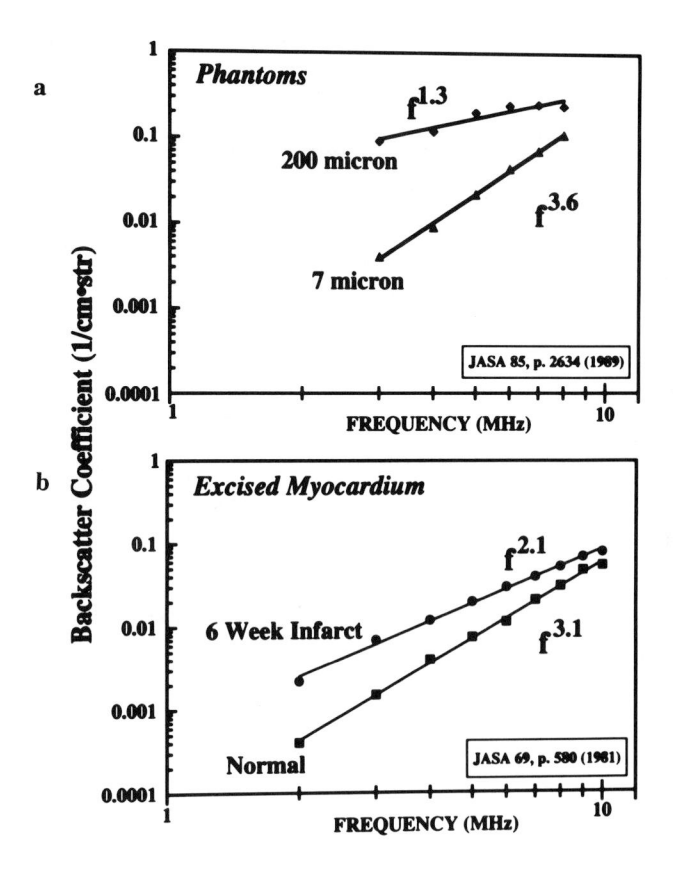

Fig. 9.8 Relationship between the size of the dominant scatter and the rate of rise of backscatter with frequency. Panel a:[29], panel b:[23]

$$\text{Sensitivity} = \frac{\text{True Positive}}{\text{Total with disease}} = \text{True Positive Fraction}$$

$$\text{Specificity} = \frac{\text{True Negative}}{\text{Total without disease}} = \text{True Negative Fraction}$$

Fig. 9.9 Definitions of sensitivity and specificity

ROC Curves for 2 Diagnostic Tests

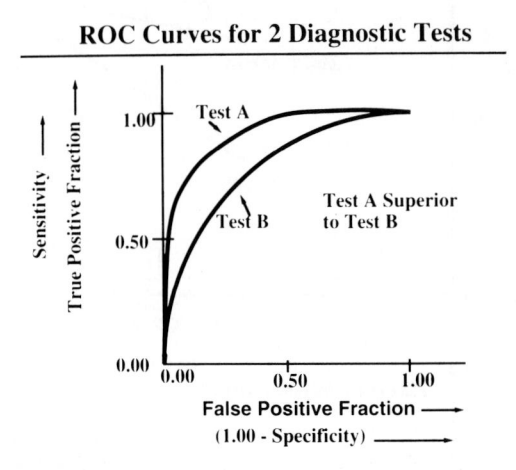

Fig. 9.10 Hypothetical Receiver Operating Characteristic (ROC) curves for two diagnostic tests. Test A is superior to test B [30]

Area Under ROC curves

Diagnostic Test	Area
"Ideal"	1.00
"Pap Smear"	0.87
"Random"	0.50

Fig. 9.11 Anticipated values for the areas under ROC curves

9. Relationships between Imaging and Myocardial TC

sulting ROC curves [34]. Results of our initial investigation suggest that patients with severe disease may have occult heart disease that was detected with tissue characterization prior to the point at which any abnormality of cardiac function is detectable using conventional echocardiography.

We first consider whether we can differentiate with the aid of tissue characterization the hearts of diabetics who have retinopathy, which might suggest relatively advanced disease, from the hearts of aged matched controls. Based on the magnitude of cyclic variation alone, the area under the ROC curve is approximately 0.85. Based on a combination of the magnitude and the time delay of cyclic variation in the septum, the area under the ROC curve rises to 0.87. Based on the magnitude and the time delay of cyclic variation measured in both the septum and the posterior wall in a single parasternal longaxis view, the area becomes approximately 0.96. These preliminary estimates of the area under ROC curves will be augmented with extensive additional considerations in a manuscript now under preparation.

Now we consider a more challenging task. We ask whether by using tissue characterization to interrogate the hearts of diabetic patients, can one predict which diabetic person has a retinopathy and which does not. Excluded from this study was any patient exhibiting evidence of heart disease by any standard clinical measure. Using the magnitude and time delay in cyclic variation in the septum one achieves an area under the ROC curve of 0.84 for this relatively difficult diagnostic test.

9.2 Indirect Applications of Tissue Characterization

Potential indirect contributions of tissue characterization take the form of improvements and extensions of current echocardiography. One example is on-line assessment of ventricular function with automatic boundary detection [35-38]. An automatic boundary detection system that is now available in a commercial echocardiographic scanner combines a previously published algorithm for determining the endocardial boundary with a real-time integrated backscatter images which serves as the input to the algorithm. Experimental tests have demonstrated conclusively that making the decision as to the location of the boundary based on the integrated backscatter signal provides a more stable and robust determination than that obtained using the conventional image. Thus a quantity that was introduced for purposes of tissue characterization provides an effective approach to the boundary detection problem, permitting real-time estimates of cardiac blood pool and of ejection fraction.

Anisotropy of Integrated Backscatter

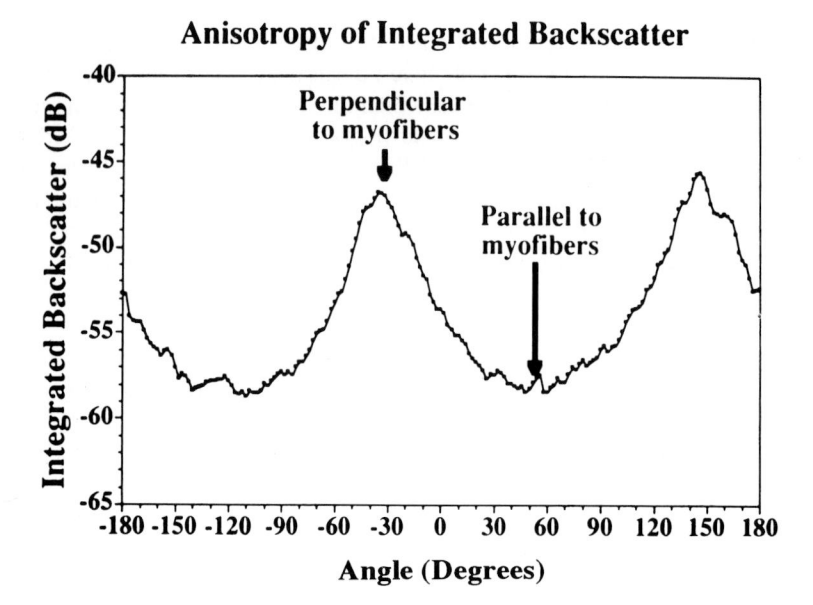

Fig. 9.12 Integrated backscatter as a function of angle, showing that backscatter is largest perpendicular to myofibers and minimum parallel to myofibers[54]

9.3 Myocardial Anisotropy

A second indirect contribution arising from tissue characterization deals with compensating for the effects arising from the anisotropy of myocardial ultrasonic properties. We have shown that values of the ultrasonic backscatter, attenuation, and speed of sound depend upon the angle between the ultrasonic beam and the local myofiber direction [39-43]. Figure 9.12 illustrates that backscatter is largest perpendicular to fiber direction and smallest parallel to fiber direction. Figure 9.13 illustrates that attenuation shows a radically different dependence on angle, with minimum attenuation perpendicular to the fibers (where the backscatter is the largest) and maximum attenuation parallel (where the backscatter is the smallest).

Figure 9.14 illustrates some effects of anisotropy in ultrasonic imaging. Imaging from views in which the ultrasonic beam propagates approximately perpendicular to the fiber direction results in relative high quality images. In contrast, imaging in views in which the ultrasonic beam is aligned approximately parallel to the fibers results in significantly degraded images, particularly in more distal zones of the image.

In order to understand how anisotropy influences echocardiographic imaging and automatic boundary detection, we consider the approximate fiber orientation encountered in a parasternal short-axis view. (See Fig. 9.15.) The weak backscatter and strong attenuation in paths near 3 o'clock and 9 o'clock, which

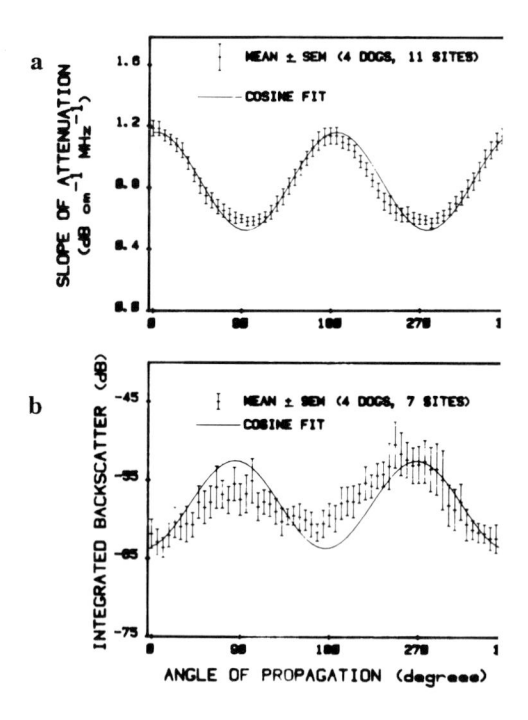

9. Relationships between Imaging and Myocardial TC   151

Fig. 9.13 Anisotropy of attenuation and of backscatter, illustrating that attenuation is maximum at angles where backscatter is minimum and the reverse. Panel a:[43], panel b:[41]

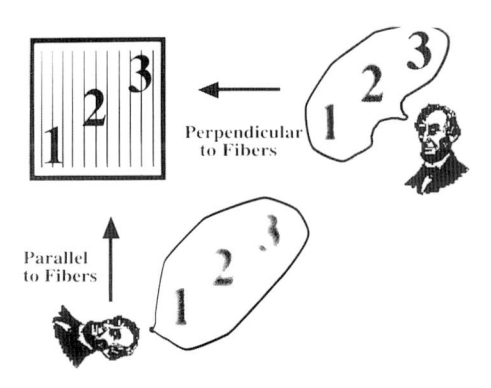

Fig. 9.14 Drawing illustrating the hypothetical effects of myocardial anisotropy of backscatter and attenuation on echocardiographic imaging

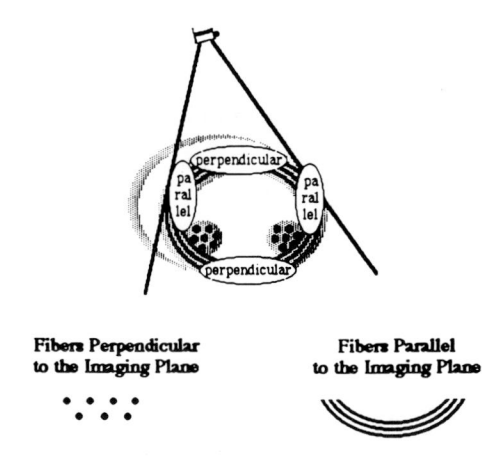

Fig. 9.15 Sketch of parasternal short axis view illustrating the approximate orientation of cardiac myofiber direction relative to the ultrasonic beam

correspond to the ultrasonic beam passing approximately parallel to the myofiber direction, often combine to produce drop-out in those regions [44, 45]. To overcome this effect of anisotropy, a novel compensation technique analogous to conventional Time Gain Compensation (TGC) has been introduced under the term Lateral Gain Compensation (LGC) [46]. With the availability of Lateral Gain Compensation, the gain can be selectively increased in pie-shaped wedges to overcome the weak backscatter and high attenuation encountered at specific angles. This is illustrated in the three panels of Fig. 9.16. In Fig. 9.16 panel a, a parasternal short axis view with the 3 o'clock and 9 o'clock regions entirely absent is shown along with the pie-shaped wedge in which the lateral gain will be increased. In panel b of Fig. 9.16 the 3 o'clock region has been restored and the automatic boundary detection algorithm has closed on that segment. For the 9 o'clock segment the lateral gain wedge has been positioned but the gain has not yet been enhanced. Panel c of Fig. 9.16 shows the final result with lateral gain enhanced in both the 3 and 9 o'clock segments and with the boundary fully closed in the entire endocardial region.

From one point of view, anisotropy is a feature of the heart that benefits from compensation with the use of lateral gain compensation. However, from a different perspective, anisotropy permits one to investigate the local transmural organization of the myofibers [47-49]. One example is illustrated in Fig. 9.17 in which the well-known progressive transmural shift that occurs from epi- through mid- to endo- in ventricular myocardium is monitored using the maximum value of integrated backscatter to determine the angle of perpendicularity. Alterations in this normal pattern might be monitored with echocardiographic tissue characterization in diseases that progressively disrupt myofibrillar architecture.

Mechanisms responsible for the attenuation and backscatter of ultrasound in

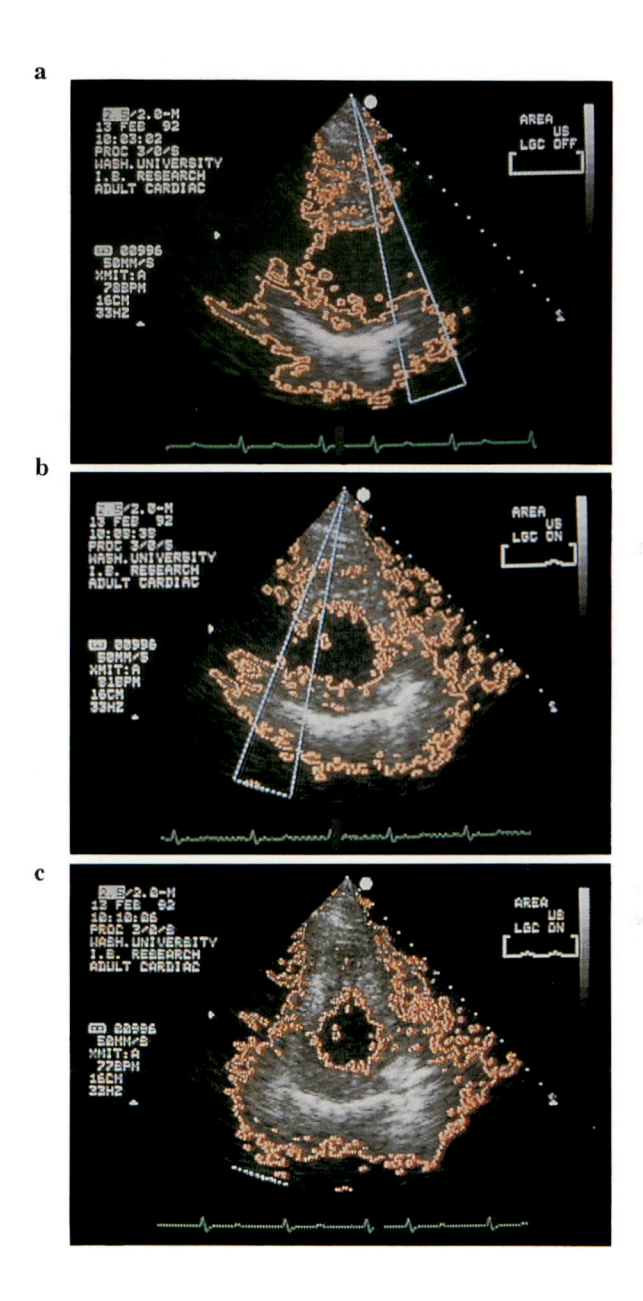

Fig. 9.16 Parasternal short axis views illustrating automatic boundary detection based on integrated backscatter and lateral gain compensation at three stages of selection and adjustment

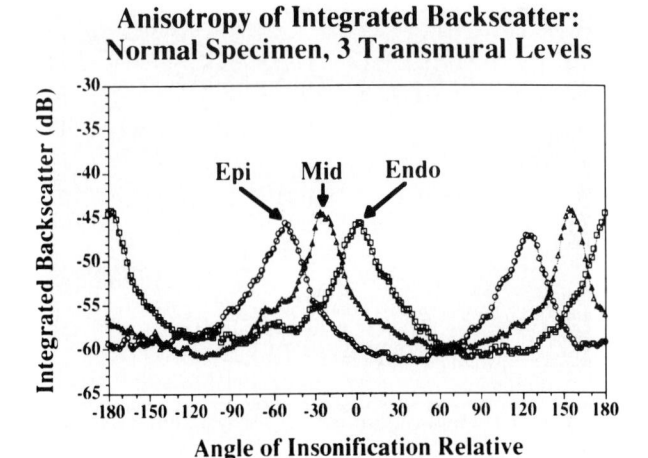

Fig. 9.17 Apparent integrated backscatter versus angle of insonification for human myocardium. The shifts in the peaks represent transmural shift of fiber orientation from epicardial to endocardial layers[47]

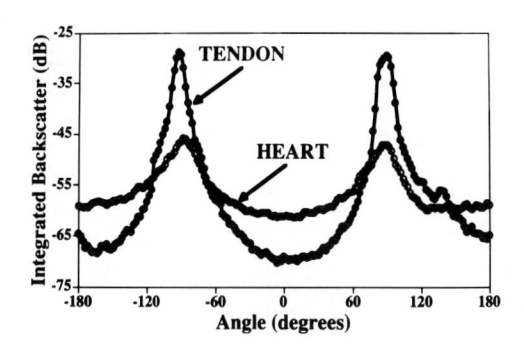

Fig. 9.18 Anisotropy of apparent integrated backscatter from human tendon and human heart[51]

soft tissue appear to be related in part to local variations in collagen content. (For a detailed discussion of this point and extensive references to pertinent literature, the reader is referred to recent article by Rose and co-authors [50].

Thus the anisotropy of integrated backscatter of tendon, a tissue with a relatively high content of collagen, might be expected to exceed that of myocardium. Data substantiating this hypothesis are presented in Fig. 9.18 [51].

9.4 Summary: Current and Future Contributions of Tissue Characterization

Investigations designed to permit quantitative ultrasonic tissue characterization based on real time backscatter imaging have already provided a basis for extending the role of two-dimensional imaging through automatic boundary detection and lateral gain compensation. Similar contributions of integrated backscatter are anticipated soon to the quantitation of studies using ultrasonic contrast agents. On the horizon is the promise of characterizing the heart itself in direct application of the techniques of ultrasonic tissue characterization.

References

1. David J. Skorton, James G. Miller, Samuel Wickline, Benico Barzilai, Steve M. Collins, and Julio E. Perez, (1991), Ultrasonic Characterization of Cardiovascular Tissue, in Cardiac Imaging: A Companion to Braunwald's Heart Disease, ed. Melvin L. Marcus, Heinrich R. Schelbert, David J. Skorton, Gerald L. Wolf, and Eugene Braunwald, 538-556, Chapter 26, W. B. Saunders Company, Philadelphia

2. Samuel A. Wickline, Julio E. Perez, and James G. Miller, Cardiovascular Tissue Characterization In Vivo, (1993), in Ultrasonic Scattering In Biological Tissue - Chapter 10, ed. Shung KK and Thieme GA, 313-345, CRC Press, Boca Raton FL

3. J. E. Perez, S. M. Collins, J. G. Miller, and D. J. Skorton, Ultrasonic Myocardial Tissue Characterization (1990), in Comparative Cardiac Imaging: Function, Flow, Anatomy, Quantitation, ed. Bruce Brundage, 403-414, Aspen Publishing Company, Gaithersburg

4. J.G. Miller, J.E. Perez, and B.E. Sobel, (1985), Ultrasonic Characterization of Myocardium, Prog Cardiovasc Dis, XXVIII, 85-110

5. Julio E. Perez, James G. Miller, Samuel A. Wickline, Mark R. Milunski, Benico Barzilai, and Burton E. Sobel, (1990), Myocardial Tissue Characterization, in Progress in Cardiology 3/1, ed. Douglas P. Zipes and Derek J. Rowlands, 83-96, Lea & Febiger, Philadelphia

6. J. Naito, T. Masuyama, J. Tanouchi, T. Mano, H. Kondo, K. Yamamoto, R. Nagano, M. Hori, M. Inoue, T. Kamada, (1994), "Analysis of Transmural Trend of Myocardial Integrated Ultrasound Backscatter for Differentiation of Hypertrophic Cardiomyopathy and Ventricular Hypertrophy Due to Hypertension", J Am Coll Cardiol, 24, 517-524

7. CM Moran, GR Sutherland, T Anderson, RA Riemersma, WN McDicken, (1994), "A Comparison of Methods Used to Calculate Ultrasonic Myocardial Backscatter in the Time Domain", Ultrasound Med Biol., 20, 543-550

8. JRTC Roelantd, GR Sutherland, S Iliceto, D Linker (Eds), Churchill Livingstone, (1992), "Tissue Characterization in Myocardial Disease" In: Cardiac Ultrasound, 419

9. F. Lattanzi, P. Bellotti, E. Picano, F. Chiarella, A. Mazzarisi, C. Melevendi, G. Forni, L. Landini, A. Distante, C. Vecchio, (1993), "Quantitative Ultrasonic Analysis of Myocardium in Patients with Thalassemia Major and Iron Overload", Circ, 87, 748-754

10. L. Landini, A. Pezzano, A. Distante, (1993), "Increased Echodensity of Transiently Asynergic Myocardium in Humans: a Novel Echocardiographic Sign of Myocardial Ischemia", J Am Coll Cardiol, 21, 199-207

11. H.U. Stempfle, C.E. Angermann, P. Kraml, A. Schutz, B.M. Kemkes, K. Theisen, (1993), "Serial Changes During Acute Cardiac Allograft Rejection: Quantitative Ultrasound Tissue Analysis Versus Myocardial Histologic Findings", J Am Coll Cardiol, 22, 310-317

12. D. Lythall, J. Bishop, R.A. Greenbaum, C.I.D. Islely, A.G. Mitchell, D.G. Gibson, M. Yacoub, (1993), "Relationship Between Myocardial Collagen on Echo Amplitude in Non Fibrotic Hearts", Eur Heart J, 14, 344-350

13. G. Gigli, F. Lattanzi, A.R. Lucarini, E. Picano, A. Genovesi-Ebert, C. Marabotti, R. Zunino, A. Mazzarisi, L. Landini, M. Iannetti, (1993), "Normal Ultrasonic Myocardial Reflectivity in Hypertensives: A Tissue Characterization Study", Hypertension, 21, 329-334

14. Lewis J. Thomas III, S.A. Wickline, J.E. Perez, B.E. Sobel, and J.G. Miller, (1986), A Real-Time Integrated Backscatter Measurement System for Quantitative Cardiac Tissue Characterization, IEEE Trans Ultrason Ferroelec Freq Contr, UFFC-33, 27-32

15. Lewis J. Thomas III, Benico Barzilai, Julio E. Perez, Burton E. Sobel, Samuel A. Wickline, and James G. Miller, (1989), Quantitative Real-Time Imaging of Myocardium Based on Ultrasonic Integrated Backscatter, IEEE Trans Ultrason Ferroelec Freq Contr, UFFC-36, 466-470

16. Julio E. Perez, James G. Miller, Samuel A. Wickline, and Mark R. Holland, (1992), Quantitative Ultrasonic Imaging: Tissue Characterization and Instantaneous Quantification of Cardiac Function, Am J Cardiol, 69, 104 - 111

17. Eric I. Madaras, Benico Barzilai, J.E. Perez, B.E. Sobel, and J.G. Miller, (1983), Changes in Myocardial Backscatter Throughout the Cardiac Cycle, Ultrasonic Imaging, 5, 229-239

18. Benico Barzilai, Eric I. Madaras, B. E. Sobel, J. G. Miller, and J. E. Perez, (1984), Effects of Myocardial Contraction on Ultrasonic Backscatter Before and After Ischemia, Am. J. Physiol. Soc. (Heart Circ. Physiol. 16), 247, 478-483

19. G.A. Mohr, Zvi Vered, Benico Barzilai, Julio E. Perez, Burton E. Sobel, and J.G. Miller, (1989), Automated Determination of the Magnitude and Time Delay ("Phase") of the Cardiac Cycle Dependent Variation of Myocardial Ultrasonic Integrated Backscatter, Ultrasonic Imaging, 11, 245-259

20. James F. Loomis Jr., Alan D. Waggoner, Kenneth B. Schechtman, James G. Miller, Burton E. Sobel, and Julio E. Perez, (1990), Ultrasonic Integrated Backscatter 2-D Imaging: Evaluation of M-Mode Guided Acquisition and Immediate Analysis in 55 Consecutive Patients, J Am Soc of Echocardiography, 3, 255-265

21. J.W. Mimbs, D. Bauwens, R.D. Cohen, M. O'Donnell, J.G. Miller, and B.E. Sobel, (1981), Effects of Myocardial Ischemia on Quantitative Ultrasonic Backscatter and Identification of Responsible Determinants, Circ, 49, 89-96

22. Benico Barzilai, Lewis J. Thomas III, Robert M. Glueck, Jeffrey E. Saffitz, Zvi Vered, Burton E. Sobel, James G. Miller, and Julio E. Perez, (1988), Detection of Remote Myocardial Infarction With Quantitative Real-Time Ultrasonic Characterization, J Am Soc of Echocardiography, 1, 179-186

23. M. O'Donnell, J.W. Mimbs, and J.G. Miller, (1981), The Relationship Between Collagen and Ultrasonic Backscatter in Myocardial Tissue, J Acoust Soc Am, 69, 580-588

24. Joel Mobley, Micha S. Feinberg, Hije M. Gussak, Christina E. Banta, Julio E. Perez, James G. Miller, (1994), "On-Line Implementation of Algorithm for Determination of Magnitude and Time Delay of cyclic Variation of Integrated Backscatter in an Echocardiographic Imager", Ultrasonic Imaging,In Press, (Abstract)

25. Zvi Vered, G.A. Mohr, Benico Barzilai, Carl J. Gessler Jr., Samuel A. Wickline, Keith A. Wear, Thomas A. Shoup, Alan N. Weiss, Burton E.

Sobel, James G. Miller, and Julio E. Perez, (1989), Ultrasound Integrated Backscatter Tissue Characterization of Remote Myocardial Infarction in Human Subjects, J Am Coll Cardiol, 13, 84-91

26. Samuel A. Wickline, Lewis J. Thomas III, J.G. Miller, B.E. Sobel, and J.E. Perez, (1986), Sensitive Detection of the Effects of Reperfusion on Myocardium by Ultrasonic Tissue Characterization With Integrated Backscatter, Circ, 74, 389-400

27. M. R. Milunski, G. A. Mohr, K. A. Wear, B. E. Sobel, J. G. Miller, and S. A. Wickline, (1989), Early Identification With Ultrasonic Integrated Backscatter of Viable But Stunned Moocardium in Dogs, J Am Coll Cardiol, 14, 462-471

28. Mark R. Milunski, Gregory A. Mohr, Julio E. Perez, Zvi Vered, Keith A. Wear, Carl J. Gessler, Burton E. Sobel, James G. Miller, and Samuel A. Wickline, (1989), Ultrasonic Tissue Characterization With Integrated Backscatter: Acute Myocardial Ischemia, Reperfusion, and Stunned Myocardium in Patients, Circ, 80, 491-503

29. Keith A. Wear, Mark R. Milunski, Samuel A. Wickline, Julio E. Perez, Burton E. Sobel, and James G. Miller, (1989), Differentiation Between Acutely Ischemic Myocardium and Zones of Completed Infarction in Dogs on the Basis of Frequency Dependent Backscatter, J Acoust Soc Am, 85, 2634-2641

30. C.E. Metz, (1978), "Basic Principles of ROC Analysis", Seminars in Nuclear Medicine, VIII, 283-298

31. Julio E. Perez, Janet B. McGill, Julio V. Santiago, Kenneth B. Schechtman, Alan D. Waggoner, James G. Miller, and Burton E. Sobel, (1992), Abnormal Myocardial Acoustic Properties in Diabetic Patients and Their Correlation with the Severity of Disease, J Am Coll Cardiol, 19, 1154-1156

32. D.J. Skorton, B. Vandenberg, (1992), "Ultrasound Tissue Characterization of the Diabetic Heart: Laboratory Curiosity or Clinical Tool?", J Am Coll Cardiol, 19, 1163-1164

33. T.J. Regan, A.B. Weiss, (1992), "Diabetic Cardiomyopathy", J Am Coll Cardiol, 19, 1165-1166

34. R. F. Wagner, K.A. Wear, J.E. Perez, J.B. McGill, K.B. Schechtman, and J.G. Miller, (1993), Multivariate Tissue Characterization: Unbiased Estimates of Asymptotic Performance Using a Finite Sample. Applications to Cardiac Ultrasonic Tissue Characterization of Diabetics and Normal Controls, Ultrasonic Imaging, 15, 156, (Abstract)

35. Julio E. Perez, Alan D. Waggoner, Benico Barzilai, H. E. Melton, James G. Miller, and Burton E. Sobel, (1992), On-Line Assessment of Ventricular Function by Automatic Boundary Detection and Ultrasonic Backscatter Imaging, J Am Coll Cardiol, 19, 313-320

36. Julio E. Perez, Steve C. Klein, David M. Prater, Carolyn E. Fraser, Hiram Cardona, Alan D. Waggoner, Mark R. Holland, James G. Miller, and Burton E. Sobel, (1992), Automated, On-Line Quantification of Left Ventricular Dimensions and Function By Echocardiography with Backscatter Imaging and Lateral Gain Compensation, Am J Cardiol, 70, 1200-1205

37. Julio E. Perez, Alan D. Waggoner, Victor G. Davila-Roman, Hiram Cardona, and James G. Miller, (1992), On-Line Quantification of Ventricular Function During Dobutamine Stress Echocardiography, European Heart Journal, 13, 1669-1676

38. Alan D. Waggoner, Benico Barzilai, James G. Miller, and Julio E. Perez, (1993), On-Line Assessment of Left Atrial Area and Function by Echocardiographic Automatic Boundary Detection, Circ, 88, 1142-1149

39. G. H. Brandenburger, J. R. Klepper, J. G. Miller, and D. L. Snyder, (1981), Effects of Anisotropy in the Ultrasonic Attenuation of Tissue on Computed Tomography, Ultrasonic Imaging, 3, 113-143

40. J.R. Klepper, G.H. Brandenburger, J.W. Mimbs, B.E. Sobel, and J.G. Miller, (1981), Application of Phase Insensitive Detection and Frequency Dependent Measurements to Computed Ultrasonic Attenuation Tomography, IEEE Trans Biomed Eng, BME-28, 186-201

41. Jack G. Mottley and J.G. Miller, (1988), Anisotropy of the Ultrasonic Backscatter of Myocardial Tissue: I. Theory and Measurements In Vitro, J Acoust Soc Am, 83, 755-761

42. E.I. Madaras, J. Perez, B.E. Sobel, J.G. Mottley, and J.G. Miller, (1988), Anisotropy of the Ultrasonic Backscatter of Myocardial Tissue: II. Measurements In Vivo, J Acoust Soc Am, 83, 762-769

43. E.I. Madaras, J. Perez, B.E. Sobel, J.G. Mottley, and J.G. Miller, (1990), Anisotropy of the Ultrasonic Attenuation in Soft Tissues: Measurements In Vitro, Jack G. Mottley and James G. Miller, J Acoust Soc Am, 88, 1203-1210

44. Alan D. Waggoner, Julio E. Perez, James G. Miller, and Burton E. Sobel, (1992), Differentiation of Normal and Ischemic Right Ventricular Myocardium With Quantitative Two Dimensional Integrated Backscatter Imaging, Ultrasound Med Biol, 18, 249-253

45. Dino Recchia, James G. Miller, and Samuel A. Wickline, (1993), Quantification of Ultrasonic Anisotropy in Normal Myocardium with Lateral Gain Compensation of Two-Dimensional Integrated Backscatter Images, Ultrasound Med Biol, 19, 497-505

46. Julio E. Perez, Steve C. Klein, David M. Prater, Carolyn E. Fraser, Hiram Cardona, Alan D. Waggoner, Mark R. Holland, James G. Miller, and Burton E. Sobel, (1992), Automated, On-Line Quantification of Left Ventricular Dimensions and Function By Echocardiography with Backscatter Imaging and Lateral Gain Compensation, Am J Cardiol, 70, 1200-1205

47. Samuel A. Wickline, Edward D. Verdonk, and James G. Miller, (1991), Three-Dimensional Characterization of Human Ventricular Myofiber Architecture By Ultrasonic Backscatter, J Clin Invest, 88, 438-446

48. Samuel A. Wickline, Edward D. Verdonk, Andrew K. Wong, Richard K. Shepard, and James G. Miller, (1992), Structural Remodeling of Human Myocardial Tissue After Infarction, Circ, 85, 259-268

49. Andrew K. Wong, Edward D. Verdonk, Brent K. Hoffmeister, James G. Miller, and Samuel A. Wickline, (1992), Detection of Unique Transmural Architecture of Human Idiopathic Cardiomyopathy By Ultrasonic Tissue Characterization, Circ, 86, 1108-1115

50. James H. Rose, Mark R. Kaufmann, Samuel A. Wickline, Christopher S. Hall, James G. Miller, (1994), "A Proposed Microscopic Elastic Wave Theory for Ultrasonic Backscatter from Myocardial Tissue", J Acoust Soc Am, J. H. Rose, M. R. Kaufmann, S.A. Wickline, C.H. Hall, James G. Miller, J Acoust Soc Am, 96, 1-13

51. B.K. Hoffmeister, A.K. Wong, E.D. Verdonk, S.A. Wickline, and J.G. Miller, (1992), Anisotropy of Ultrasonic Backscatter from Human Tendon Compared to That from Normal Human Myocardium, .Proc. IEEE Ultrasonics Symposium, 91CH3079-1, 1127-1131

52. J.G. Miller, J.E. Perez, Jack G. Mottley, Eric I. Madaras, Patrick H. Johnston, Earl D. Blodgett, Lewis J. Thomas III, and B.E. Sobel, (1983), Myocardial Tissue Characterization: An Approach Based on Quantitative Backscatter and Attenuation, Proc. IEEE Ultrasonics Symposium, Atlanta, 83 CH 1947-1, 782-793

53. Lewis J. Thomas III, Benico Barzilai, Julio E. Perez, Burton E. Sobel, Samuel A. Wickline, and James G. Miller, (1989), Quantitative Real-Time Imaging of Myocardium Based on Ultrasonic Integrated Backscatter, IEEE Trans Ultrason Ferroelec Freq Contr, UFFC-36, 466-470

54. Edward D. Verdonk, S. A. Wickline, and J. G. Miller, (1990), Quantification of the Anisotropy of Ultrasonic Quasi-Longitudinal Velocity in Normal

Human and Canine Myocardium With Comparison to Anisotropy of Integrated Backscatter, Proc. IEEE Ultrasonics Symposium, 1990, Honolulu, 90CH2938-9, 1349-1352

Discussion

OPHIR: Thank you, Professor Miller. We have some time for questions.

HILL: There's one interesting bit of physics that comes out of your talk. You commented that the attenuation coefficient is highest along the fibers where the scattering is lowest. That essentially agrees with observations that we've made some time ago, but we never understood the physics. Do you understand it?

MILLER: Well, only partly. First, it certainly suggests that backscatter is playing a negligible role in the total attenuation, since obviously they go in opposite directions. We have long felt that the language needs to be refined slightly, in that we tend to say that attenuation is the sum of scattering plus absorption. There's nothing wrong with that statement, but we need to be careful that we recognize that there are two types of scattering that we need to deal with. An initially longitudinal wave comes in and scatters to another longitudinal wave, and those are the waves that Bob Waag showed us earlier this morning, and they have their own angle dependencies. They scatter at essentially all angles, though sometimes preferentially at some more than others. And that's longitudinal to longitudinal mode scattering. However, at the very same time, the original longitudinal mode wave hits an inhomogeneity and mode converts to a shear wave, and there's now a shear wave scattered away, but it dies out very quickly–so quickly that it looks like absorption; in fact, it is absorption. So it's really a matter of language or semantics. It is true that attenuation is the sum of scattering plus absorption. But there are multiple forms of scattering: longitudinal to longitudinal, which is the most common kind and the kind that Dr. Waag has talked to us about, and which I have talked about a bit for backscatter; longitudinal to shear, which then decays very quickly; and in fact other modes that are equally real that essentially amount to longitudinal modes going into nonpropagating modes which we associate with heat. In a certain sense, they are the mechanisms of absorption. It's our contention that if we could understand all of those forms of scattering, that we would be able to answer Kit Hill's question. We've made only a tiny bit of progress in that direction, but we are actively doing measurements with transverse waves in order to try to map out some of these effects. They're not ready yet for sharing, but in a year or two perhaps.

QUESTION: Frequency dependency of the backscattering...along the fiber, because you characterized the backscatter frequency dependency sometime proportional to the [INAUDIBLE] low frequency, or three power low frequency. Something like that.

MILLER: It's a very good question. The only frequency dependencies that I showed in this talk were for backscatter. And they were specifically confined to that angle. But we have some measurements not yet reported of the angle dependence of the frequency dependence of backscatter. It's an awful lot of words to put together. The angle dependence of the frequency dependence of backscatter. And there is an angle dependence to the frequency dependence, and we even have the beginnings of a theory to explain it. But I'd kind of like

9. Relationships between Imaging and Myocardial TC

to save that for an original presentation at another meeting. So let me just say that we have some measurements and we have data every two degrees around 360 degrees. So it's a very reliable data set, and there is a substantial effect as a function of angle in terms of the frequency dependence. I will share this with you. The issue of what the frequency dependence is very pertinent to beating heart studies, because even if you try to stay perpendicular to the heart wall, as the heart moves, because there's not only contraction but many other movements, you're never strictly perpendicular. And you might not have been able to see it in my slides, and I deliberately went quickly through it, but the frequency dependence when you have excised heart and you can be very perpendicular is more like frequency to the 3 or 3.5. But the frequency dependence in an open-chested dog, where we could see the heart but it was moving, was less than that. As you correctly realize, that is definitely related to the fact that that's coming off angle a little bit, and that's enough to lower the apparent frequency dependence. The frequency dependence when you're strictly perpendicular is the fastest, and as you come off, as you imply, it goes down. And so if you look at our in vitro work and compare it with our own in vivo work, you'll see two different slopes, but they're not wrong. In vitro, we can be very perpendicular. In vivo we can only average to be perpendicular, and we see a slower rise. It still permits us to differentiate old scar from acute ischemia, but the absolute slopes, as you correctly guessed, are a little different. We definitely reported that, and I'll actually publish soon the frequency dependence of the angle dependence of the backscattering. So thank you for the question.

TANAKA: Thank you very much for a very informative, interesting presentation you have given us. As far as integrated backscattering is concerned, it is very effective to know various characteristics of the heart walls. I have been looking at your works for many years, and I have been very much impressed. There is one point about which I would like to ask. In integrated backscattering, what is the echo source here? Of course, they are speckle echoes, but in case of the muscle of the heart walls, I think other parameters are also included.

MILLER:...the echoes internal to the tissue. This really goes back to the early paper that Professor Dunn and his student Scott Fields published many years ago, pointing to the role of collagen in both attenuation and backscatter, in this case. It is known that there is a collagen sheath around the myocytes and around bundles of myocytes, and so our hypothesis is that the contrast that produces the local echoes is between the myocyte and its bounding sheath of collagen. Of course, the echoes that we see with integrated backscatter are the same echoes seen with conventional processing. If you ask what advantages, if any, of integrated backscatter might be, in a way it comes back to discussions we've already had this morning. If you look at the speckle size in a conventional image, the conventional speckle is shaped like a banana. But the speckle size in integrated backscatter is shaped more like a spherical object, or an object whose dimensions are the same. The reason is that we deliberately average in the time or depth domain to make this speckle size approximately spherical or box shaped, as opposed to banana shaped. That modest amount of additional averaging helps to stabilize the signal. Of course, it's measuring the same thing that any other processing measures, but we deliberately blur along the time dimension to make the speckle instead of this banana-shaped thing, which is thin in this direction but long in this direction, we lengthen it this way as well. And that provides extra stability, and the proof of that is that in the Hewlett-Packard Instrument in the developmental stage we could throw a switch and run boundary detection on the conventional image on the screen, or on the integrated backscatter image. It just doesn't work very well on the conventional image, and it works rather well on integrated backscatter. And the key to that is just having a quick processor to estimate integrated backscatter in real time. And that's done even in that commercial machine rather crudely, and yet it works adequately. But off-line, we can do it in an increasingly sophisticated way. And so the trade-off, and our first talk this morning really hit upon it–tissue characterization can only work, as Kit Hill has told us, if it happens fast enough that the clinician can get feedback. And the speed for cardiology–I heard it said and I understand that in radiology real time is about one second–but in cardiology, real time is about a thirtieth of a second, because you've got to follow things. And so it has to run very quickly, even though it's a very compromised, very quick, approximate calculation. And we have somewhat better ways, but it's a trade-off of time and computing power. Thank you for your question.

TANAKA: There is one additional point related to integrated backscattering. One factor affecting integrated backscattering is sample volume, or spatial resolution. As we have stated, the mean value, the average, is necessary, for if the sample volume is very large the mean value will be lower, and if the beam is very thin and sample area is very small, then the variation might be quite large. What is the optimum sample volume of spatial resolution, especially on the ultrasound diagnostic devices. The thickness of the beam is well controlled, but it might have the effect upon this if the beam is widened, and the angle dependency would play a big role. So in terms of spatial dependency, spatial resolution, what is the effect?

MILLER: We have done experimental studies of the effect of the integration time in integrated backscatter. The second set of images that you saw on the videotape corresponded first to the Hewlett-Packard real-time commercial imager, second to the integrated backscatter. That integrated backscatter runs with an integration time of 3.2 microseconds. And 3.2 microseconds is approximately four times more integration time than the conventional image, which is ever so roughly .8 microseconds. So we settled on that number not by theoretical considerations, but by looking at the images, by looking at the values, by looking at what the outcome was. It's a little unsophisticated, but it's very practical. We also routinely report the results on a dB, on a logarithmic scale, again for reasons which are practical as opposed to theoretically sound. And there are a number of compromises that have been built in, which we would not do in laboratory processing, but which make the system work in a practical way, and again addressing the question that Kit Hill said, that we have to have something that turns around quickly enough to be useful. Of course, in the lateral direction the resolution for integrated backscatter is the same as in conventional imaging. So whatever the lateral beam focusing characteristics are in a 128 element array in this case, we get the same benefits, because integrated backscatter does not degrade that, but it does deliberately degrade the time or depth resolution, and right now that's set at four times the Hewlett-Packard machine does, and the others are probably pretty comparable. So it's only four times, and it makes the speckle spots uniform in size instead of banana shaped.

TANAKA: Thank you very much. Can I ask one more question? At the WFUMB conference, I asked this point already to you. By looking at your investigation at your laboratory, the amplitude of the curve of integrated backscattering has been used quite frequently in conventional works. But more recently, I have seen papers dealing with the gradient of the curves rather than the amplitude. What is the reason why you have been focusing upon the gradient?

MILLER: ...deals with the fact that given the integrated backscatter wave form, there are a number of features of it that you might analyze. The most obvious feature is simply the magnitude. How many dB does it go down? Five dB. Another feature that I just mentioned was the time delay relative to the systolic interval. And that is very important, but we haven't talked about it at this meeting. Still another parameter which Dr. Tanaka knows very well and is referring to is the rate at which the curve falls. If you have a young, healthy heart, it falls very quickly. It really comes down like that. If you have a not-so-healthy heart, it takes its time getting down. And so the first derivative as a function of time—the time rate of change of the integrated backscatter signal—is an interesting parameter and definitely contains diagnostic information. And as Dr. Tanaka has said, we have and others have published some papers showing that. It's a local measure, it's approximately a local measure of contractile function. Because you can measure it at any place in the image that you want. In addition to that, the whole curve is also interesting because in acute ischemia, integrated backscatter goes up three or four or five dB, and in old infarct it goes up 10 dB. The group at Osaka, Toru Masuyama and his colleagues, have a very good technique for absolute calibration of that curve, which works in patients by referencing it to the backscatter from blood in the ventricular cavity. And they are ahead of us, way ahead of us, in St. Louis in that regard. Working with the engineers from Fujitsu, they have an absolutely calibrated machine. Now I should quickly say to my colleagues, when we say absolutely calibrated, here's how they avoid the problems of not knowing attenuation through the chest wall or path. They measure the blood pool right next to the heart. So they have the known scattering from the red cells in blood very close to where they want to measure, and that removes a very difficult problem. We all know, and we heard several presentations say, that if you don't know the attenuation along the way, then you don't know the backscatter. Well, my friends at Osaka have solved that problem, and the Fujitsu people and the cardiologists at Osaka are to be congratulated. We at St. Louis have tried to do what they have tried to do, but we have not yet succeeded. I think we will, but we haven't yet; they're ahead of us. So absolute backscatter, magnitude of backscatter, cyclic variation, time delay of backscatter cyclic variation, and then how fast it comes down –all are valuable, they're all useful parameters, I believe, and many have been used by us and others. Thank you.

OKAWAI: Three years ago you visited Japan and you gave us a very stimulating presentation at that time. I would like to thank you again. My question is as follows. As one of the applications of backscattering, the automatic boundary detection was pointed out by you. And I would like to ask you about the precision of that. What is the precision accuracy of that, for example? Conventionally, when we try to detect boundaries, M-mode method or B-mode method was used. Therefore, we used our eyes to make a judgment. In your method of backscattering, which was introduced today to us, what is your accuracy of precision? And in relation to that, in normal endocardium, between two cases with healthy endocardium and abnormal endocardium, are there any differences in terms of the precision of backscattering, between the healthy ones and abnormal, pathological ones?

MILLER:...the reliability of automatic boundary detection and the parameters derived from it, for example, ejection fraction. I should quickly say that although I have much to thank my friends at Hewlett-Packard for, because they have provided me with equipment and help, I do not want it to seem as though I am an advocate of their company to the exclusion of any other. They are a commercial company that makes one form of automatic boundary detection. I have the good fortune that it turns out that their boundary detection runs on our algorithm. I personally am delighted at the fact that it performs. My cardiologist colleagues have done careful comparisons in St. Louis. More importantly, cardiologists who do not have any connection with the work have also done comparisons, and those are the ones to pay attention to. And there have been some good studies showing that it is a good technique, a useful technique, but not a perfect technique. It has limitations. Like many things in cardiography, it takes a skilled operator, because there are gain settings that are always done. Time gain compensation, lateral gain compensation, transmit level, other things that the echocardiographer routinely does in every study. Those do influence the outcome. But for the people here who do not work in cardiology and who do not have a chance to realize it, our cardiologist friends–Dr. Nitta and many others could tell you– that good echocardiography requires a very skilled hand, and the cardiologist chooses a few representative beats out of many terribly bad beats. Beats where respiration is right, and then makes an intelligent judgment. So given the limitations of echocardiography and the dependence that it has on operator skill, this automatic boundary detection is useful, helpful, limited and not perfect, has many things to be improved. The other commercial companies have their own approaches to the problems, which I think are also valid. I show it more to share with the audience the fact that in the short run, the efforts that many people in this room have contributed to for 20 years have at least led to a commercial product that's selling, and that's important. How can we justify being here if something that's commercializable doesn't come out of this work? So thanks to the work of many people, including some work from St. Louis, there is at least one product that runs on tissue characterization. There is inside of every Hewlett-Packard machines that runs automatic boundary detection, inside is a real-time integrated backscatter image. But the person who buys it isn't permitted to see that. It's just what's used to make the decision. But that's locked out from the user. But you've got one in there if you have one.

OPHIR: This is the end of this session. Thank you very much.

Part III

ACOUSTIC ANATOMY

Chapter 10

Acoustic Microscope for the Tissue Characterization in Medicine and Biology

10.1 Introduction

10.1.1 Purpose of Development of a Scanning Acoustic Microscope and the Measurement Method of Acoustic Properties of Biological Tissues

In order to develop a new ultrasonics application to medicine, it is necessary to use effectively the information which the ultrasound possesses and to know the acoustic properties of the tissue in the normal and diseased states.

As suggested by Fig.10.1, biological tissue is composed of a variety of elements having different sizes and properties and it is acoustically heterogeneous. Generally, two approaches can be considered for the investigation of the structural features of such media. In the method illustrated by Fig.10.1(a), the average values, or the summation of properties of various tissue elements and the effects of scattering, interference and diffraction are observed because the width of the acoustic beam is larger than the dimensions of the individual tissue elements. The data obtained with such a method yields macroscopic bulk properties.

On the other hand, the individual structural elements can be observed, as suggested in Fig.10.1(b), when the acoustic beam width is small enough to distinguish among the individual tissue elements such as muscle fiber, epithelial and interstitial tissues. The data obtained with such method yields microscopic properties. This methodology is described further.

---

*M. Tanaka,M.D,Ph.D., H. Okawai,Ph.D., N. Chubachi,Ph.D., R. Suganuma,B.S., and K. Honda,Ph.D.

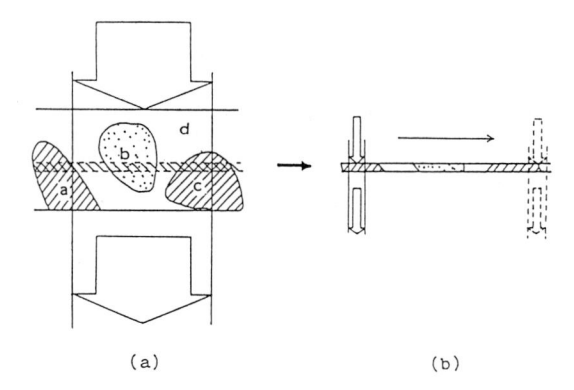

Fig. 10.1 Two methods for measuring acoustic propagation properties of biological tissues. (a) Object is the average or sum of properties of numerous tissue elements by using the acoustic beam wider than the elements, viz., macroscopic properties. (b) Object is the properties of individual tissue elements, viz., microscopic properties

Table 10.1 Sequence of acoustic microscope developments

Year	Author	Description
1936	S. Sokolov	Ultrsound microscope.
1959	F. Dunn, W. Fry	Ultrasonic absorption microscope
1969	A. Korpel, P. Desmares	Rapid sampling of acoustic holograms by laser scanning technique.
1972	A Korpel, L.W. Kessler, M. Ahmed	Brag diffraction sampling of sound field.
1972	L.W. Kessler	Scanning laser acoustic microscope.
1973	R.A. Lemons, C.F. Quate	Mechanicaly scanned acoustic microscope.

10.1.2 A Brief Acoustic Microscope History

The stages of acoustic microscope proposals are shown in Table 10.1. Sokolov was the first to suggest an acoustic microscope in 1936[1] and demonstrated his method at 1 MHz. In 1959 Dunn and Fry operated a 12MHz ultrasonic absorption microscope using a thermocouple probe as the detector[2]. Development at the microscope level was realized by Kessler with a scanning laser acoustic microscope, SLAM, in 1972[3] and by Lemmons and Quate with a mechanically scanned acoustic microscope, SAM, in 1973[4].

In the SLAM, the specimen is placed between a substrate and a coverslip, with the spacing distance determine by a spacer of known dimension so that precise values of specimen thickness are obtained. This method is superior for determining the speed of propagation and the attenuation constant of the sample[5],[6].

10. Acoustic Microscope for TC

In the SAM, an acoustic lens, or concave transducer, produces a narrow beam so that high lateral resolution is achieved.

Generally, the higher the ultrasonic frequency, the greater will be the lateral resolution, though greater difficulty will be experienced in quantitative measurement of the acoustic properties. In fact, imaging in the gigahertz frequency range has been realized[8], but the methodology is not appropriate for quantitative measurement. On the other hand, quantitative measurement of properties is routinely performed by the V-Z curve method[9], though it is not applicable to high resolution imaging. A method dealing with specimens with thickness around 10μm, in order to measure acoustic properties of tissue elements and to image their distribution, has been developed[7]. A major problem of this method is that the actual thickness of the specimen is not easily obtained because the tissue is not placed between two plates as in the SLAM, and it may vary in shape over the extent of measurement. Method to solve this problem, and its effects are described herein.

10.2 Acoustic Microscope System in the Frequency Range 100-200 MHz

Figure10.2 is a block diagram of the SAM comprising; [1] acoustic elements, [2] a mechanical scanner, [3] analogue signal processor, [4] an image display unit, and [5] an image processor. Figure10.3 shows an assembly of the system and Fig10.4 show unit [1] and [2]

10.2.1 Ultrasonic Transducer

The ultrasonic transducer are illustrated in Fig.10.5. The upper one is an acoustic focusing element comprises a ZnO piezoelectric transducer with a sapphire lens. The aperture half angle is 30° and focuses the acoustic beam with a 1/2 power beam width lateral resolution from approximately 6 μm at 200 MHz to 12μm at 100 MHz, in water at 20 °C. The lower one is a plane type transducer which is 6 mm in diameter wider than the scan width.

10.2.2 Mechanical Scanner

The mechanical scanner is shown in Fig.10.4. The upper acoustic element is fabricated on a plate which is vibrated at 60 Hz in the X-direction by the electro magnetic method. The rod, on which both the tissue and its holder are mounted, is slided reciprocally in the Y-direction in 8 sec by the motor. Thus, two dimensional scanning is performed to produce C-mode images. The scanning width is variable as 2 mm × 2 mm, 1 mm × 1 mm, or 0.5 mm × 0.5mm.

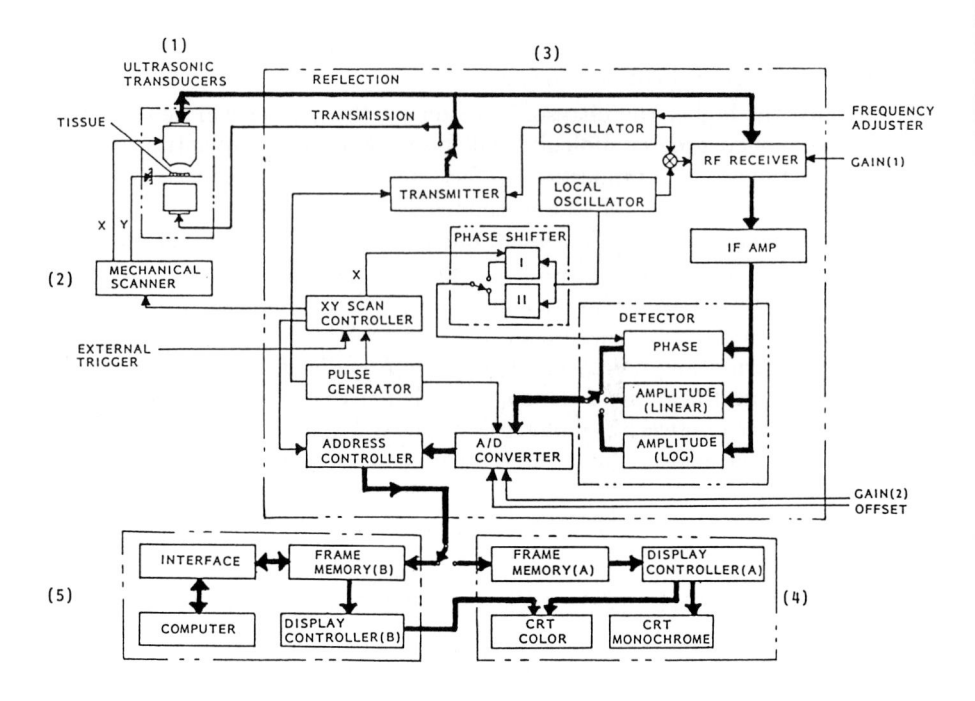

Fig. 10.2 Block diagram of the mechanically scanned acoustic microscope system. (1) Ultrasonic transducers, (2) mechanical scanner, (3) analogue signal processor, (4) imaging part, (5) image signal processor

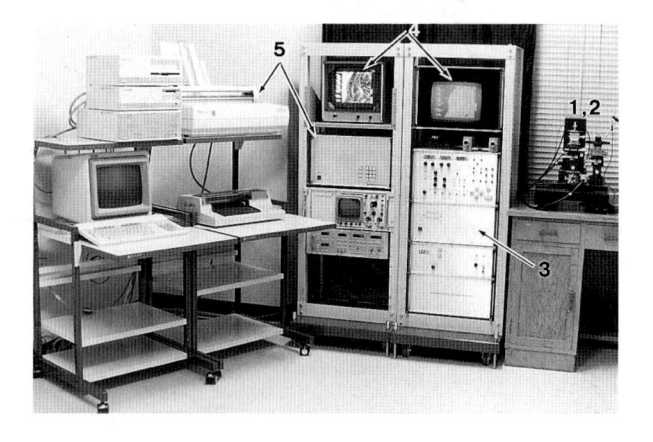

Fig. 10.3 The microscope system. View of the complete system: the numbers in the figure corresponds to those of block diagram of the mechanically scanned acoustic microscope system

10. Acoustic Microscope for TC   175

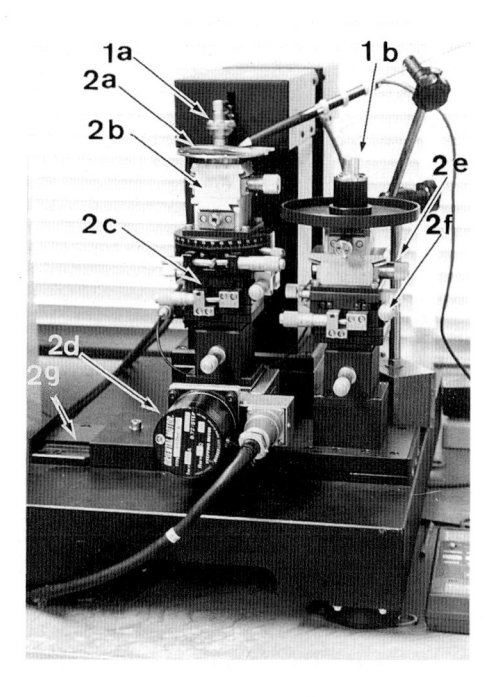

Fig. 10.4 The microscope system. View of the ultrasonic transducers and mechanical scanner: 1a: focusing type transducer, 1b: plane type transducer, 2a: specimen holder, 2b: adjuster for alignment of the specimen holder, 2c: adjuster for XYZ positions of the specimen holder, 2d: stepping motor for Y scan, 2e: adjuster for alignment of the plane type transducer, 2f: adjuster for XYZ position of the plane type transducer, 2g: slide for reflection and transmission methods

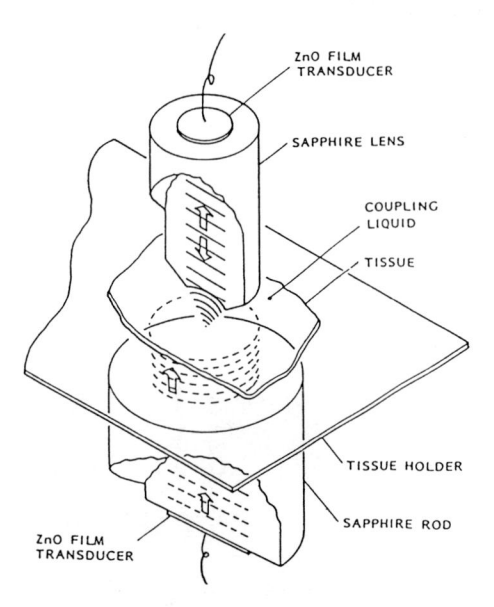

Fig. 10.5 The ultrasonic transducers. The sound wave is irradiated from upper transducer in the reflection method and from the lower one in the transmission method

10.2.3 Analogue Signal Processor

This unit comprises a pulse generator, a XY scanning controller, an acoustic transmitter and receiver, a detector, an A/D converter, and a XY address controller.

The pulse generator supplies 500-1000 μs pulses to the acoustic element. The received signal is sent to the RF receiver, IF amplifier and to the detector. The amplitude detector detects signals linearly in the range of 20 dB. Figure 10.6(a) shows amplitude acoustic images for which the color scale means that the nearer to red, the larger the ultrasonic attenuation, and black corresponds to 0 dB, i.e., tissue is not present. The phase detector compares the phase of the received signal with the output of the phase detector II and produces $\sin\phi$, where ϕ is the phase difference. The level of $\sin\phi$ is adjusted as negative for the wave transmitted when the tissue absent (level a in the figure) and positive for the tissue present (level b). The phase for tissue present gains $\Delta\phi$ over the case for the tissue absent, as shown in Fig. 10.7. The phase image Fig. 10.6(b) presents $\sin\Delta\phi$ for the condition that black corresponds to level a and the other color corresponds to b were red is the maximum. The precise value of the sound speed is obtained as

$$C_s = \frac{1}{\frac{1}{C_w} - \frac{\Delta\phi}{2\pi f \delta}} \qquad (10.1)$$

where C_w, f, and δ are the sound speed of the coupling medium, the ultrasonic

10. Acoustic Microscope for TC   177

OPTICAL IMAGE   AMPLITUDE IMAGE   PHASE IMAGE (130 MHz)
                                      ACOUSTIC IMAGE

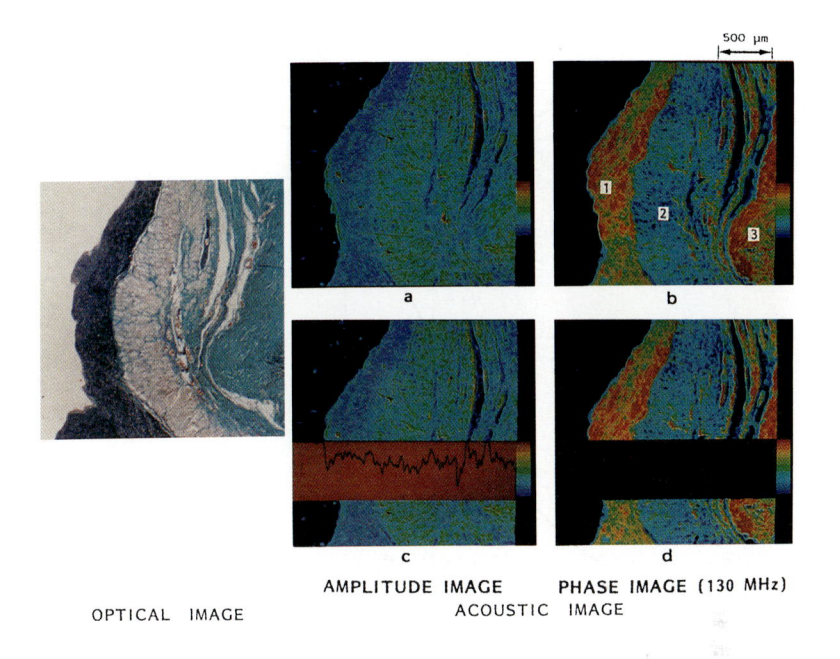

Fig. 10.6   SAM images (frequency: 130 MHz, scanning width: 2mm). (a) Amplitude, (b) Phase, (c) and (d) A-mode traces overlapped with the SAM images. The specimen is an example of the infarcted myocardium which was formalin fixed and once paraffin embedded: 1: thickened portion in the endocardium, 2: degenerated portion, 3: scar portion

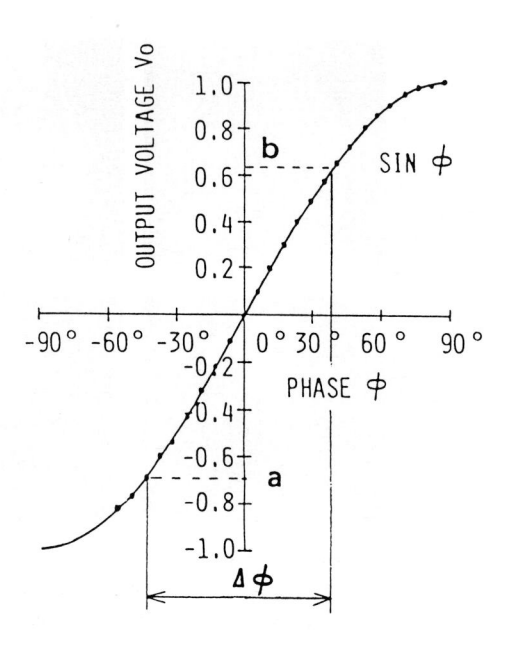

Fig. 10.7 Output of phase detection. The output of the phase detection is $\sin\phi$ of which ϕ is subtraction of the phases between received the signal and the output of phase shifter II. The phase shift $\Delta\phi$ is the difference in which the levels a, b are of the received waves with and without tissue present, respectively

10. Acoustic Microscope for TC   179

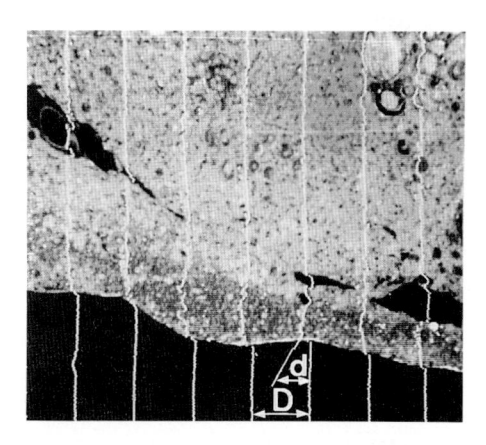

Fig. 10.8 SAM images: interferogram, at 170 MHz and 2mm in scanning width, overlapped with amplitude images

frequency and the thickness of the specimen (two times of thickness for the reflection method) respectively.

When $\Delta\phi$ is larger than 180°, the phase image is not appropriate, so the interferogram is used. If the phase out put of the phase shifter II is delayed in synchrony with the upper acoustic element moving in the X direction, the phase difference between the received signal and the reference signal becomes larger with increase of X-position of the upper element. The interferogram is made by tracing the pulses, which occur at times when the phase difference becomes 0, 2π, 4π, $\cdots\cdots$, with the Y scan, as shown in Fig.10.8. In Fig.10.8(a), the amount of fringe shift d and interval D correspond to one period and phase shift $\Delta\phi$, respectively.

Here, the sound speed of the specimen is given as,

$$C_s = \frac{f\delta C_w}{f\delta - NC_w} \qquad (10.2)$$

where N is the relative shift of fringes, $N = d/D$.

The value at the time of gating, which is the output of the pulse oscillator, is A/D converted in 6 bits. The error of coding of V is, around center of full scale,

$$\Delta V = 20[\log(33/63) - \log(32/63)] = 0.267 \text{dB}$$

The coding step of the phase shift is 1.8 for the condition that input range is $-\sin 90°$ to $+\sin 90°$. This value corresponds to 2 m/s for a 10 μm thick tissue specimen, in the reflection mode at 130 MHz.

The amplitude and phase data are sorted in a two dimensional address in the XY address controller.

10.2.4 Imaging Part

The two dimensionally sorted data are stored in a 6 bit frame memory (A) and displayed in color scales of 10-16 shades, as shown in Fig.10.6. The detailed level of the data in any X scanning line is displayed in the A-mode, as in Fig.10.6(c)-(d).

10.2.5 Image Signal Processor

This comprises the image memory (B) and a computer (HP310) processes the image signal.

10.3 Measurement Method of Acoustic Properties of Thin Tissue Specimens

10.3.1 Non-contact Method

Knowledge of the thickness of the sample is essential for determination of sound speed and attenuation of a specimen. Generally, the thickness of a specimen is determined by placing it between two solid plates with parallel contact surfaces of known separation, such as a micrometer for solid materials.

Such a contact method cannot be employed for tissue specimens of the order of 10 μm in thickness because important details of the contact of the tissue specimen to solid plates are unknown. Moreover, contact of solid plates to soft tissue can change the specimen thickness and, therefore, the determined acoustic properties. In addition, the acoustic beam must be scanned to detect the acoustic properties for each position on the inhomogeneous material and the value of the specimen thickness should be obtained under the condition that the specimen is immersed within coupling liquid. For these reasons, specimen thickness should be measured by a non-contact method.

Lee et al. studied a non-contact measuring method for thin mental films by analyzing the interference of sound waves produced on interfaces spaced the thickness of the film[10]. A non-contact method, which utilizes such interference of sound waves is developed here for determining specimen thickness essential for obtaining the acoustic properties at the microscopic level of thin biological tissue specimens, as illustrated in Fig.10.1(b)[11],[12]. Tissue specimens less than 10 μm in thickness are considered because(1) this is the thickness generally examined with the light microscopy in histology and pathology laboratories and (2) tissue elements can be resolved and examined acoustically individually, without confusion with other neighboring elements.

10.3.2 Principle of Measuring Method

For biological tissue, which are assumed to have the characteristic behavior of fluid media as regards acoustic phenomena, the sound speed C, density ρ,

10. Acoustic Microscope for TC

adiabatic volume elasticity K and adiabatic volume compressibility β are related by

$$C = \sqrt{\frac{K}{\rho}} = \frac{1}{\sqrt{\rho\beta}} \qquad (10.3)$$

With respect to ultrasonic attenuation A in biological tissues, it is considered that absorption α and scattering σ contribute as $A = \alpha + \sigma$. For classical absorption wherein only the viscous mechanism need be considered, the absorption coefficient is given by

$$\alpha = \frac{2\pi^2 f^2}{\rho C^3} \frac{4\eta}{3} \qquad (10.4)$$

where f, and η are ultrasonic frequency, and shear viscosity, respectively[13]. The observed absorption is, however, greater than the value predicted by eq.(10.4) and it increases nearly linearly with frequency rather than quadratically. This is believed to be due to the delay between stress and strain arising when sound propagates in a relaxing medium. It is reported that the absorption phenomena is influenced by the protein (macromolecular) content of the tissue medium [13]–[18]. In the frequency range 1-10MHz, the attenuation and the absorption in the tissue increase approximately with $f^{1-1.2}$ [19),20].

It is not known whether diseased tissue exhibit this same behavior, and it is, therefore, necessary that both attenuation and sound speed be available. Because of the general lack of attenuation, sound speed, absorption, scattering data, and their frequency dependencies, the data obtained by ultrasonic microscopy cannot be compared easily with data reported in the literature and it is for these reasons that this new method is considered.

With the arrangement of the tissue specimen of approximately 10 μm in thickness mounted on the glass slide, a three layered structure of coupling medium, tissue section, and glass slide is obtained. The thickness corresponds to 1/2 to 1/3 wavelength so that the principle of interference can be employed. Consequently, the reflection coefficient frequency characteristics exhibits a unique pattern, depending upon tissue thickness.

As illustrated in Fig.10.9, the wave is focused at the surface of the glass slide. As regards the contributions of the various wave components, it is known that the received signal is mainly the round trip wave y_1 and the wave reflected from the surface of the tissue y_2 [11),12].

The far surface of the specimen is considered to be flat due to its intimate contact with the glass substrate. The near surface, however, which is free from constraint, may not be flat microscopically as suggested in Fig.10.10. The level of amplitude and phase of the above areas are suggested by a and c in Fig.10.10.

The maximum value of the thickness is considered as A in Fig.10.10. This simplifies the wave components y_1, y_2, and as

$$y_1 = \frac{2Z_2}{Z_1 + Z_2} \frac{Z_3 - Z_2}{Z_3 + Z_2} \frac{Z_2 - Z_1}{Z_2 + Z_1} \exp(-2(\alpha_2 + j\beta_2)l) \qquad (10.5)$$

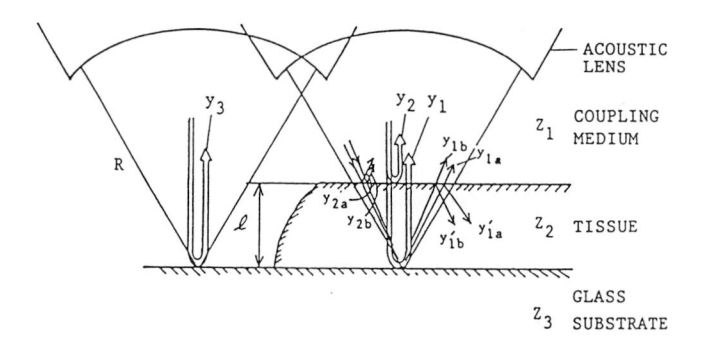

Fig. 10.9 Wave components of the reflected wave. Because of focused wave, the reflected wave is composed of the normal incident wave and oblique incident wave. Effectively, the round trip wave y_1 and the wave reflected from upper surface of a specimen y_2, which can be calculated for normal incident waves, mainly contribute. The wave y_3, reference wave, is reflected from a glass substrate without tissue present. $R= 1.25$mm, $l= 10$ μm

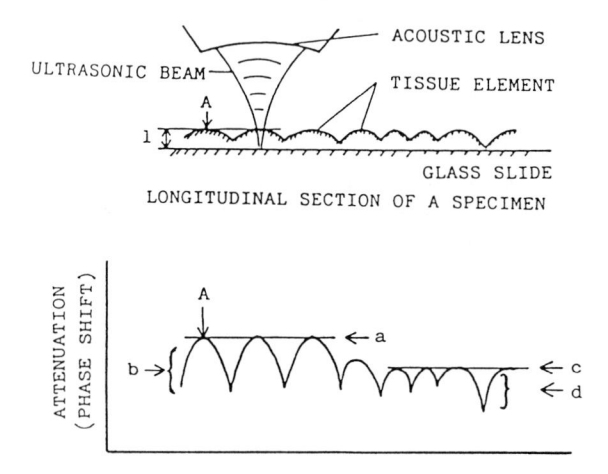

Fig. 10.10 Hypothetical model to discuss the relationship between longitudinal section of a specimen and the beam width. The front surface of portion at maximum attenuation or phase shift of an A-mode in a tissue element is parallel to the glass substrate. Thickness of this portion is adopted for the specimen thickness

10. Acoustic Microscope for TC

Table 10.2 Acoustic constants and nations

	Coupling medium (water 20 °C)	Sample (tissue)	Sample mounting (glass slide)
Sound speed (m/s)	$C_w = 1483$ *	C_s	
Density (g/cm³)	0.9983	$\rho = 1.06$	
Acoustic impedance (Kg/m²s)	$Z_1 = 1.480 \times 10^6$	$Z_2 = \rho C_s$	$Z_3 = 17 \times 10^6$
Phase constant (radian/m)	β_1	β_2	
Attenuation constant (neper/m)		α_2	

* Greenspan and Tschiegg, 1958

$$y_2 = m_1 \frac{Z_2 - Z_1}{Z_2 + Z_1} \tag{10.6}$$

$$y_3 = \frac{Z_3 - Z_1}{Z_3 + Z_1} \tag{10.7}$$

where y_3 is the wave reflected from the tissue holder without the tissue present and m_1 is a correction factor the for the decrease of the acoustic pressure wave amplitude due to off-focus positioning, and determined as

$$m_1 = 10^{-m/20} \tag{10.8}$$

$$m = 0.004 \exp(0.029F) \quad [\text{dB}] \tag{10.9}$$

(F is the frequency in MHz), for specimen thickness of approximately 10 μm [11].

The acoustic constants used here are listed in Table 10.2. Consequently, the attenuation L and the phase shift ϕ are given by,

$$L = -\left\{ 10 \log(y_1 + y_2)^2 - 10 \log\left(\frac{Z_3 - Z_2}{Z_3 + Z_2}\right)^2 \right\} \tag{10.10}$$

$$\phi = \arg((y_1 + y_2)/y_3) \tag{10.11}$$

where the acoustic impedance is considered to be a real number. The ultrasonic attenuation of water at 20°C, the coupling medium, is approximately 0.18 dB at 200 MHz, as determined from

$$\alpha/f^2 = 25.3 \times 10^{-17} \quad \text{s}^2/\text{cm} \quad [21]$$

The value of the attenuation arising here is approximately 2 dB so that the contribution to the attenuation of water can be neglected.

Figure 10.11 shows the calculation of the attenuation and the phase shift obtained by substituting the values from Table 10.3 into eqs.(10.6)-(10.11), where

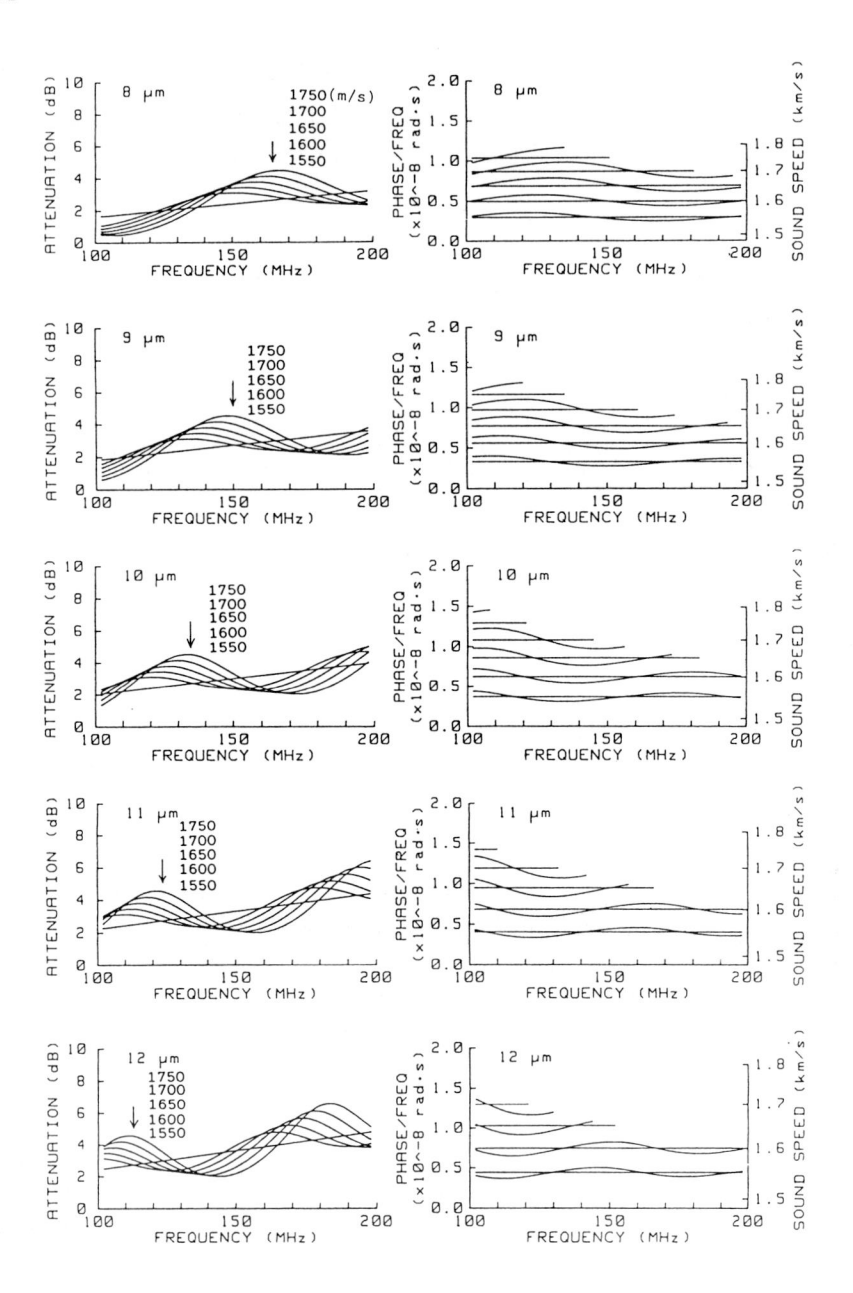

Fig. 10.11 Calculation of attenuation and phase shift of the wave from the model of Fig.10.9. Thickness = 8 - 12 μm. The composite wave of $y_1 + y_2$ undulate the traces, although the wave y_1 itself varies linearly with frequency. These straight lines strictly reveal attenuation and sound speed

Table 10.3 Values substituted to calculate attenuation and phase shift of a tissue model

attenuation constant	$\alpha_2 = 1.151 \times 10^{-4} f$ (neper/m) (f:Hz)
	(from 100 dB/mm at 100 MHz, assumed slope of unity)
sound speed	C_s = 1550 to 1750 m/s
	(assumed non-dispersive)
specimen thickness	ℓ = 8 to 12 μm

the sound speed of water, the coupling medium, is 1483 m/s at 20°C [22], under the assumption that the frequency exponent in the attenuation is unity and the sound speed in non-dispersive. If the patterns obtained experimentally are the same as the traces of Fig. 10.11, then the values of specimen thickness, attenuation constant and sound speed can be determined.

10.3.3 Tissue Preparation and Measurement

The sample tissue was formalin fixed, paraffin embedded and sectioned with a microtome to a thickness of 10 μm for the acoustic microscope and to a thickness of 3μm for the light microscope. As shown in Fig. 10.12, it is known experimentally that the attenuation increases 50 to 100 %, and that the sound speed increases 20 to 40 m/s as a result of formalin fixation.

Additionally, these properties increase 100 % and 40 m/s, respectively, due to paraffin embedding. This increase is believed arise because of the reaction of formallin and amine, dehydration, and cross linking to stabilize macromolecular structure [23]. Paraffin embedding produces shrinking of tissue [24] and it is considered that as the amount of free water decreases, the speed, absorption and scattering of sound increases.

The following equation was used for determining the attenuation L, instead of eq. (10.11),

$$L = -(10\log(y_1 + y_2)^2 - 10\log y_3^2) \qquad (10.12)$$

The following losses are included in eq. (10.12), 1) mismatch between the coupling medium and the tissue specimen, 2) attenuation in the pathway, and 3) reflection loss at the glass substrate. The first loss is negligible as it contribute less than 1 % to y_1. The error introduced by the third loss is a maximum of approximately 10 % at 100 MHz.

Consequently, the attenuation listed in Table 10.4 shows a maximum absorption, scattering and reflection loss of 10 %.

10.3.4 Experimental Result for Tissues

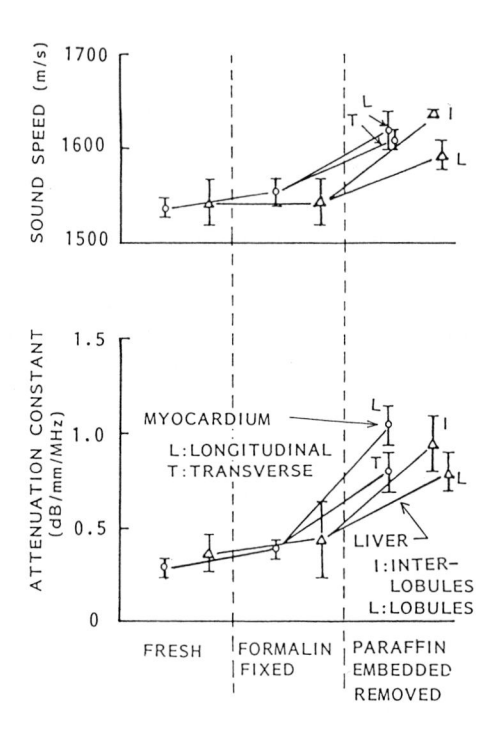

Fig. 10.12 Experimental results of the effect of formalin fixing and paraffin embedding to acoustic properties of tissues. Five tissue sections of myocardium and lives of rat were used at each state. Fresh: the tissues were sectioned at frozen state and set on glass substrates. Formalin fixed: the specimens above were immersed with formalin in one day. Paraffin embedded and removed: the tissue blocks of same organ were embedded and sectioned, and the paraffin was removed just before measuring

10. Acoustic Microscope for TC 187

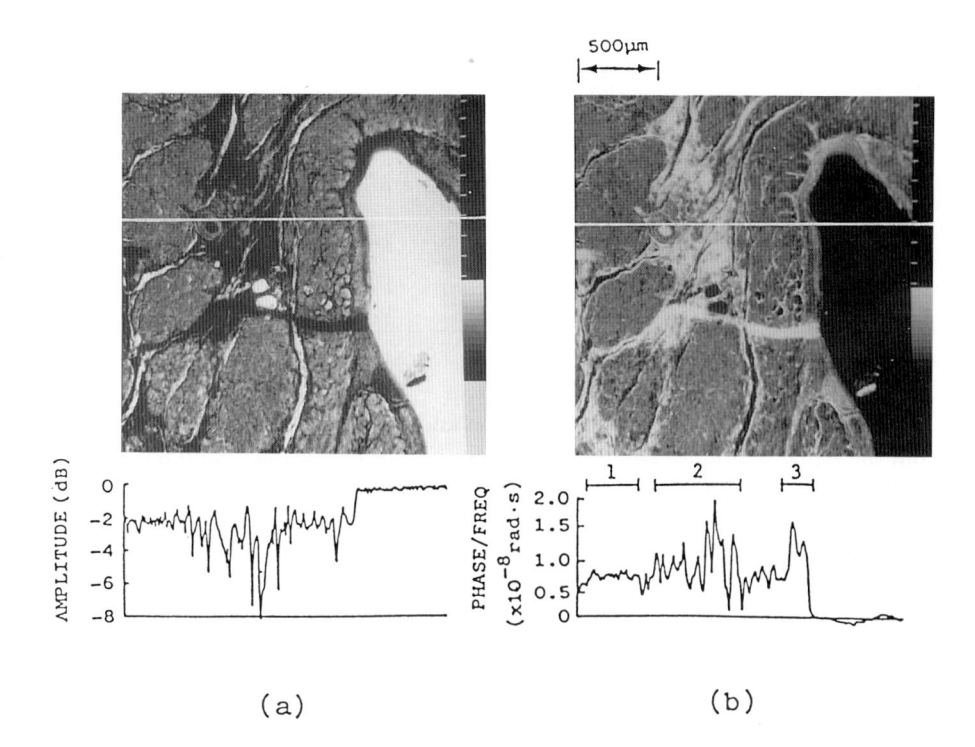

(a)

(b)

Fig. 10.13 SAM images of (a) amplitude, (b) phase and A-mode profiles for a myocardium sample, around 10 μm thick at 20°C. A-mode profiles show amplitude and phase profiles along the white lines. On the plot, the area under bar 1: normal cardiac muscles area, 2: collagenous changed area, 3: elastic changed area

Table 10.4 Experimental data, attenuation constant and sound speed of the tissue sections, around 10 μm thick, cut from three infarcted myocardia of human origin, once paraffin embedded, in the frequency range 100 to 200 MHz (20°C)

			attenuation (dB/mm) 100MHz	200MHz	slope	sound speed(m/s)
(a)						
	Mn	×	125	250	1	1600
	Md	△	80	190	1.2	1580
	Fc	○	150	300	1	1710
	Fe	○	80	190	1.2	1700
(b)						
	Mn	×	100	200	1	1590
	Md	△	70	150	1.1	1540
	Fc	○	150	300	1	1770
	Fe	○	100	250	1.3	1650
(c)						
	Mn	×	80	200	1.3	1630
	Fc	○	100	240	1.3	1710

Figure 10.13 shows images of amplitude and phase. The levels were obtained from the A-mode profiles along the white lines shown. The maximum levels, over small portions, i.e., portions parallel to the glass substrate, were from several samples. The data thus obtained are shown in Fig.10.14. Thus, the levels of both attenuation and phase shift exhibit variation. It is considered that the variation in thickness, though the specimens were sectioned with a microtome.

Thus, the thicknesses were determined by comparing the frequency characteristic patterns from the experiments with the calculations. In specimen (b) of Fig.10.13, the attenuation curve has a maximum in the frequency range 120 to 135 MHz, a minimum in the range 150 to 170 MHz. The phase curve has a maximum in the neighborhood of 100 MHz, a minimum in the range 135 to 145 MHz. Thus, the thickness can be judged to be 10 μm. The thicknesses of samples (a) and (b) were determined by the same procedure. The actual attenuation and speed were determined from the straight lines shown in Fig.10.14 and the results are listed in Table 10.4.

Judging from Fig.10.14, and the entries of Table 10.4, it is considered that the frequency exponent of the attenuation is 1 to 1.3 ($\alpha = Af^n$) and that the velocity is relatively non-dispersive. As regards dispersion of the sound speed in biological tissue, some observations have been reported[25]-[27]. Kremkau et al observed sound speed dispersion in human brain to be approximately 1 m/s/MHz in the frequency range 1-5 MHz[25]. However, the experimentally determined graph in their paper also shows the speed increasing linearly with the logarithm of frequency. If such a dependence also occurs in the range 100-200 MHz, the sound speed would increase approximately 2 m/s. That is, in this frequency range, dispersion would not be observable because the measurement method is not accurate to 2 m/s.

10. Acoustic Microscope for TC

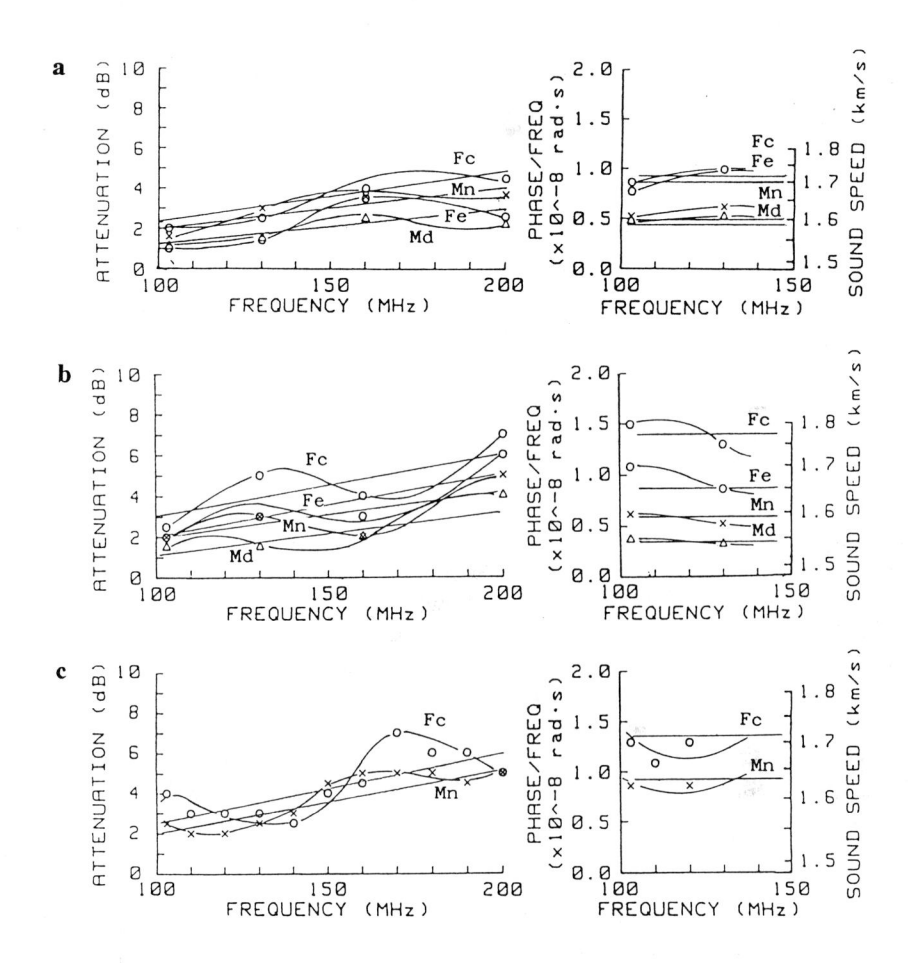

Fig. 10.14 Experimental data of attenuation and phase shift in the tissue of three infarcted myocardia (a,b,c). Mn: normal cardiac muscles areas (x), Md: degenerated cardiac muscles area (△), Fc: collagenous changed area (○), Fe: elastic changed area (○)

Thus, once the data is obtained experimentally, the specimen thickness is determined and the attenuation constant and sound speed can be obtained.

10.4 Quantitative Two-dimensional Display Method of Acoustic Properties

The original images of amplitude and phase shift need to be sophisticated to reveal attenuation and speed. The frequency characteristics of amplitude and phase of the reflection wave obtained experimentally have patterns similar to those of Fig.10.11, so that the interest is put in optimum frequency to make two-dimensional quantitative display.

As shown in Fig.10.11, the optimum frequency for attenuation occurs at a point where the curves of different sound speed intersect. For example, the optimum frequency is 145 MHz for 8 μm thickness, and 130 and 175 MHz for 9μm. In the neighborhood of these frequencies, the amplitude of the waves transmitted though tissue elements having the same attenuation constant takes on the same levels independent of sound speed. The level is a little different from the actual attenuation shown in the straight line traces, so that appropriate compensation would make the levels at the optimum frequency display the actual attenuation constant. The optimum frequency thus obtained is in the range 120 to 160 MHz.

In this frequency range, the error arising in selecting a certain frequency under the condition that the frequency exponent is in the range 1 to 1.3, is within 7 %. The error is produced by variation of the crossing points of the curves having different speeds, approximately 10 % at maximum.

Regarding phase shift, for thicknesses less than 10 μm, the phase shift is greater than that for the round trip wave. On the other hand, for thicknesses greater than 10 μm, the phase shift is less. Putting this tendency to attenuation, it became known that, in the range 120 to 130 MHz, phase shift curves correspond to correct scales for speed in the graphs determined under the assumption that the tissue thickness is 10 μm. Consequently, selection of 120 or 130 MHz produces the error within 10 m/s, in principal as shown in Fig.10.15. Thus, the frequency exponent n in the attenuation $\alpha = Af^n$ is 1 to 1.3 and the dispersion of sound speed is less than 2 m/s. Therefore, in order to obtain images of attenuation constant and sound speed, n was set to 1, the attenuation constant was normalized by f^1, and the sound speed was assumed to be constant in the frequency range 100 to 200 MHz. By doing this, the attenuation constant presented in dB/mm/MHz and the sound speed in m/s become independent of frequency.

The acoustic beam is scanned two-dimensionally to display spatial distributions of the tissue element properties. The color scales of ten shades, shown in Table 10.5, are chosen to reveal simultaneously the attenuation constant and sound speed of each pixel, i.e., sampling point, quantitatively in the two-dimensions. The advantage of this arrangement is that the color scales can be applied in any

10. Acoustic Microscope for TC

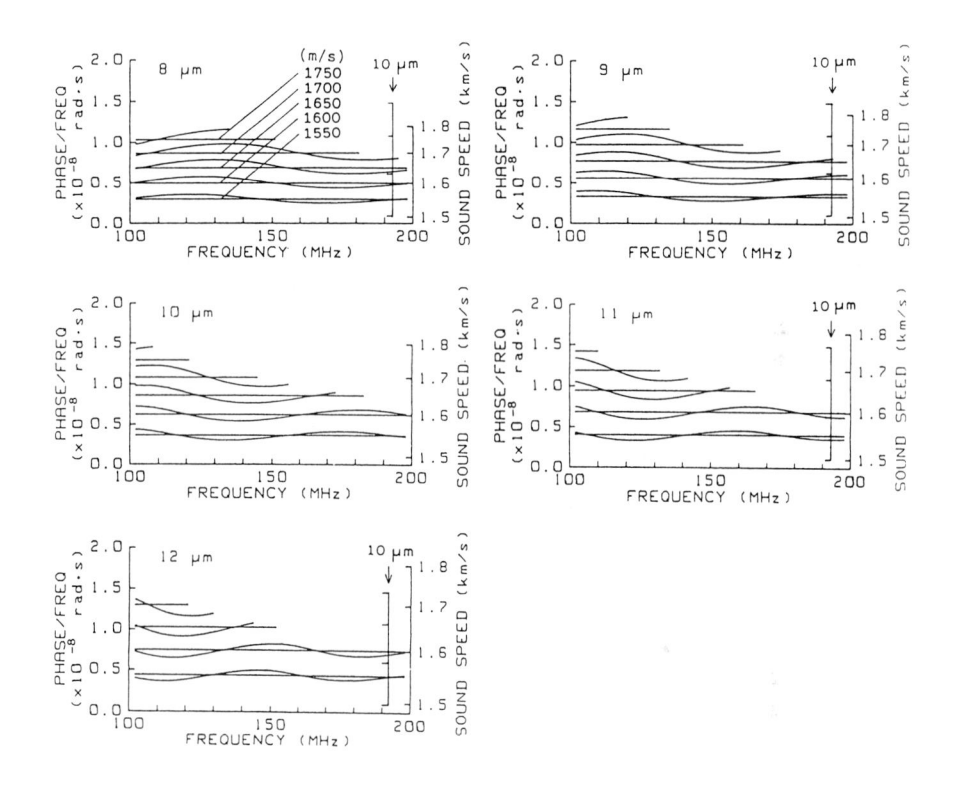

Fig. 10.15 Relationship between phase shift and sound speed. The scale labeled 10 μm shows sound speed under the assumption that the thickness of a tissue is 10 μm

Table 10.5 Values of sound speed and attenuation constant. Each values are divided (a) in 30 m/s and 0.25 dB/mm/MHz, respectively (b) 20 m/s and 0.2 dB/mm/MHz, respectively

a

SPEED		ATTENUATION	
units:m/s		units:dB/mm/MHz	
red	(≥ 1765)	red	(≥ 2.375)
magenta	(1750 ± 15)	magenta	(2.25 ± 0.125)
orange	(1720 ± 15)	orange	(2.0 ± 0.125)
brown	(1690 ± 15)	brown	(1.75 ± 0.125)
yellow	(1660 ± 15)	yellow	(1.5 ± 0.125)
green	(1630 ± 15)	green	(1.25 ± 0.125)
olive green	(1600 ± 15)	olive green	(1.0 ± 0.125)
cyan	(1570 ± 15)	cyan	(0.75 ± 0.125)
royal blue	(1540 ± 15)	royal blue	(0.5 ± 0.125)
blue	(1510 ± 15)	blue	(0.25 ± 0.125)
black	(<1495)	black	(<0.125)

b

red	(≥ 1690)	red	(≥ 1.9)
magenta	(1680 ± 10)	magenta	(1.8 ± 0.1)
orange	(1660 ± 10)	orange	(1.6 ± 0.1)
brown	(1640 ± 10)	brown	(1.4 ± 0.1)
yellow	(1620 ± 10)	yellow	(1.2 ± 0.1)
green	(1600 ± 10)	green	(1.0 ± 0.1)
olive green	(1580 ± 10)	olive green	(0.8 ± 0.1)
cyan	(1560 ± 10)	cyan	(0.6 ± 0.1)
royal blue	(1540 ± 10)	royal blue	(0.4 ± 0.1)
blue	(1520 ± 10)	blue	($0.2 ^{+0.1}_{-0}$)
black	(<1510)	black	(<0.2)

10. Acoustic Microscope for TC

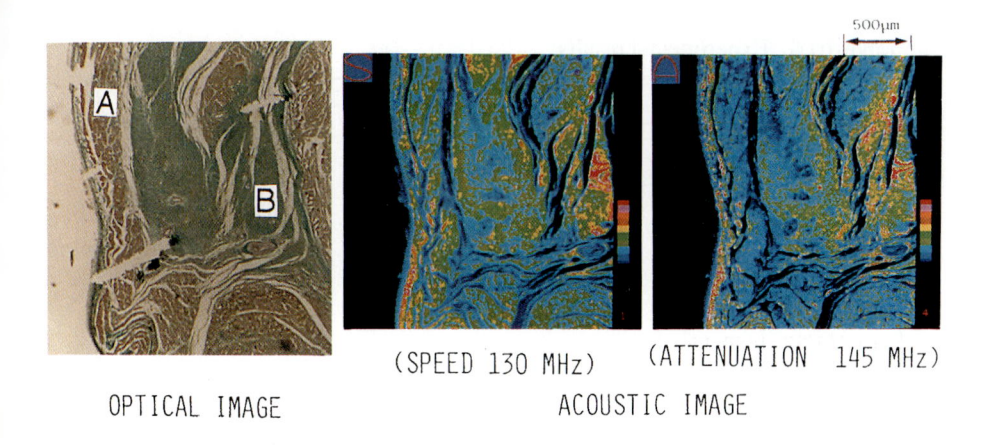

Fig. 10.16 Examples of SAM images (sound speed and attenuation) compared with the light microscopic image (optical image)

frequency in the range 100 to 200 MHz.

Figure 10.16 shows an example of the quantitative two dimensional images of attenuation and sound speed of the myocardium thus obtained, the specimen was taken from an experimental animal (dog) with myocardial infarction. The thickness of the specimen was determined to be 11 μm with the non-contact procedure. The left side picture is an optical image of the infarcted myocardial tissue stained by the Elastica Masson's trichrom method. Middle and right side pictures are acoustic images of sound speed and attenuation respectively. The acoustic properties in the area A which is myocardial area show about 1600 to 1630 m/s in sound speed and about 0.75 dB/mm/MHz in attenuation constant.

In the collagenous changed area B, the sound speed is in the rage of 1630 to 1720 m/s and the attenuation constant 1.25 to 2.0 dB/mm/MHz.

From these results, it can be said that the local acoustic property in the biological tissue is two-dimensionally displayed and estimated quantitatively by the use of our acoustic microscopic system.

Although in section 10.3, thickness was determined by comparing frequency characteristic patterns, computer processing for the entire process is now being studied as the next step in the development of the system [28],[29].

10.5 Data Acquisition of Acoustic Properties of Biological Tissue

Acoustic microscopic images which are framed by the following two procedures; 1) adopting the image processing described in section 10.4 and the color shading as shown in Table 10.5, and Fig.10.16. 2) adopting 16 color shades,

Table 10.6 Experimental results

prep.	ROIs or MGs	region	attenuation (dB/mm/MHz)	speed (m/s)
brain 1.rat 1tumor	2MGs	cerebral cortex tumor necrosis bleeding	0.9-1.4 0.6-0.9 1.1-1.6	1560-1590 1500-1550 1500-1670
colon 1. human 1CAN	2MGs	normal lamina propria intestinares folliculus lymphaticus cancerous folliculus lymphaticus	0.8 1.1 0.4 0.8	1640 1740 1600 1640
heart 1. human 1NR 3MIs	49ROIs 3MGs	normal longitudinal transvers degenerated elastic changed area mixed area of collagenous and elastic fibe	1.07±0.12 0.81±0.10 0.71±0.20 1.36±0.19 1.04±0.07 1.28±0.16	1623±23 1610±9 1569±23 1729±39 1670±25 1704±50
2. human 3CMs	3MGs	normal collagenous changed area		1700-1900
3. dog 2MIs		normal longitudinal collagenous changed area	0.7±1.1 0.9±2.1	1590-1670 1590-1760
kidney 1. human 1NR		urinary tuble malpighian corpuscle collagenous area (loose)	0.6-1.1 0.7-1.3 0.7-0.8	1560-1610 1560-1610 1560-1610
liver 1. rat 3NRs	3ROIs	lobules interlobules	0.7-0.9 0.8-1.1	1580-1610 1640
2. rat 1YS	3MGs	normal cancerous(YS)	0.5-0.7 0.2-0.3	1590-1630 1530-1610
3. human 1CIR	4MGs	pseudolobules collagenous changed area	0.9-1.6 0.7-2.5 (more)	1590-1680 1560-1780 (more)
mammary gland 1. human 1NR	3MGs	normal epidermis corium cancerous area	0.2-0.6 1.2-2.5 (more) 0.2-0.3	1560-1610 1630-1780 (more) 1560-1580
stomach 1. human 1CAN	12MGs	nomal lamina propria subglandular region of the lamina propria muscularis mucosae blood vesels in the submucosa muscularis externa edema mucosa canserous mucosa(dense) (loose) dense collagenous muscularis externa	0.6-1.4 0.4-0.8 0.4-0.8 0.9-1.6 0.4-0.8 <0.3 1.1-2.5 (more) 0.4-0.6 1.1-2.5 (more)	1620-1730 1560-1580 1620-1640 1620-1670 1530-1580 1500-1580 1620-1780 (more) 1590-1640 1620-1780 (more)
tendon 1. dog 1NR	1MG	transvers	1.9-2.5 (more)	1800-1900

NR: normal; MI: myocardial infaction; CM: cardiomyopathy; CAN: cancer
YS: Yoshida sarcoma; CIR: cirrhosis; MG: acoustic micrograph
temprature 20-22 C

though not a strictly quantitative display, without the above image processing, as shown in Fig.10.6,

Table 10.6 shows results obtained by following three processes; 1) to read from A-mode profiles, shown in Fig.10.13, for example, 2) to read by color scales in two-dimensional image, as shown in Table10.5, the data obtained by 1) and 2) are near the average and standard deviation, 3) to calculate strictly average and standard deviation by using A-mode profiles, where prep in Table 10.6 shows the number of examples, ROI is the area of interest which was measured in the A-mode 50 to 100 μm, and MGs is the number of micrographs of which more than 5 areas of 50 to 100 μm in diameter were measured.

References

1. Sokolov S., The ultrasonic microscope. Akadema Nauk SSSR, Doklady 64, 333-335, 1949.

2. Dunn F. and Fry W., Ultrasonic absorption microscope. J. Acoust. Soc. Am. 31, 632-633, 1959.

3. Kessler L.W., Palermo P. R. and Korpel A., Practical high resolution acoustic microscopy. Acoustic holography 4, 51-71 (Wade G. ed.) Plenum Press, 1972.

4. Lemons R.A. and Quate C.F., A scanning acoustic microscope. Ultrason. Symp. Proc. IEEE, 18-20, 1973.

5. Goss S.A. and O'Brin W.D.Jr., Direct ultrasonic velocity measurements of mammalian collagen threads. J. Acoust. Soc. Am. 65(2), 507-511, 1979.

6. O'Brin W.D.Jr. Olerud J. Shung K.K., et al., Quantitative acoustic assessment of wound maturation with acoustic microscopy. J. Acoust. Soc. Am 69(2), 575-579, 1981.

7. Tervola K.M.U., Gummer M.A., Erdman J.W.Jr., and O'Brien W.D.Jr., Ultrasonic attenuation and velocity properties in rat liver as a function of fat concentration: A study at 100 MHz using a scanning laser acoustic microscope. J.Acoust. Soc. Am. 77(1), 307-313,1985.

8. Quate C.F., Acoustic microscopy(Recollections). IEEE Trans. Sonics. Ultrason. SU- 32, 132-135, 1985.

9. J.Kushibiki and N.Chubachi, Material characterization by line-focused-beam acoustic microscope. IEEE Trans. Sonics, Ultrason. SU-32(2), 189-212, 1985.

10. Lee C.C. Tsai C.S. and Cheng X., Complete characterization of thin-and thick-film materials using wideband reflection acoustic microscopy. IEEE Trans. Sonics. Ultrason. SU-32, 248-258, 1985.

11. Okawai H., Tanaka M., Chubachi N. and Kushibiki J., Non-contact simultaneous measurement of thickness and acoustic properties of a biological tissue using focused wave in a scanning acoustic microscope. Proc. 7th Symp. Ultrason. Elec., Kyoto, Jpn., J. Appl. Phys. Supple. 26-1, 52-54, 1987.

12. Okawai H., Tanaka M., and Dunn F, Non-contact acoustic method for the simultaneous measurement of thickness and acoustic properties of biological tissues. Ultrasonics Vol.28(6), 401-410, November 1990.

13. Fry W.J. and Dunn F., Analysis and experimental methods in biological research. (Physical techniques in biological research 4. Chapter 6.) 261-394. Academic Press (New York), 1962.

14. Dunn F., Edmonds P.D., and Fry W.J., Absorption and dispersion of ultrasound in biological media. (Biological Engineering. Schwan H.P. ed. Intra-university electronics 9, Chapter3) (McGraw-Hill), 205-332, 1969.

15. Pauley H. and Schwan H.P., Mechanism of absorption of ultrasound in liver tissue. J. Acoust. Soc. Am. 50(2), 692-699, 1971.

16. Bamber J.C., Fry M.J., Hill C.R., and Dunn F., Ultrasonic attenuation and backscattring by mammalian organs as function of time after excision. Ultrasound in Med. & Biol., 3. 15-20, 1977.

17. Johnston R.L., Goss S.A., Maynard V., Brady J.K., Frizell L.A., O'Braien W.D.Jr., and Dunn F., Elements of tissue characterization (Part I. Ultrasonic propagation properties.) National Beaurou standards, Spec. Publ. 525, 19-27, 1979.

18. Goss S.A., Johnston R.L., Maynard V., Nider L., Frizzell L.A., O'Brien W.D.Jr., and Dunn F., Elements of tissue characterization (Part ||. Ultrasonic parameter measurements.) National Beaurou standards, Spec. Publ. 525, 43-51,1979.

19. Chivers R.C. and Hill C.R., Ultrasonic attenuation in human tissue. Ultrasound in Med. & Biol. 2, 25-29, 1975.

20. Goss S.A., Frizzell L.A., and Dunn F., Ultraconic absorption and attenuation in mammalian tissues. Ultrasound in Med. & Biol. 5, 181-186, 1979.

21. Pinkerton J.M.M., The absorption of ultrasonic waves in liquids and its relation to molecular constitution. Proc. Phys. Soc. B62, 129-141, 1949.

22. Greenspan M. and Tschiegg C.E., Tables of the speed of sound in water. J. Acoust. Son. Am. 31, 75-76, 1958.

23. Watanabe T., Manual for preparing the pathological tissue specimen. Japanese pathological society 36-50, 1981(in Japanese).

24. Mori H. and Ishiguro K., Tissue shrinking during embedding. 11th report of meeting for tissue preparation technique 1977(in Japanese).

25. Kremakaw F.W., Barnes R.W. and MacGraw C.P., Ultrasonic attenuation and propagation speed in normal brain. J. Acoust. Soc. Am. 70. 29-38, 1981.

26. Garstensen E.L. and Schwan H.P., Acoustic properties of hemoglobin solutions. J. Acoust. Soc. Am. 31, 305-301, 1959.

27. O'Donnell M., Jaynes E.T. and Miller J.G., General relationships between ultrasonic attenuation and dispersion. J. Acoust. Soc. Am. 63, 1935-1937, 1978.

28. Okawai H., Tanaka M., Dunn F., Chubachi N. and Honda K., Quantitative display of acoustic properties of the biological tissue elements. Acoustical Imaging 17, 193- 201, 1989.

29. Okawai H., Ohtsuki S., Tanaka M. and Dunn F., A frequency sweeped type non- contact acoustic property measurement method. JSUM proceeding 56, 267-268, May 1990.

Discussion

THIJSSEN: Thank you for your paper. The paper is now open for discussion. Professor Hill.

HILL: Thank you for your paper. If I understand your results correctly, you were measuring attenuation in tissue of about 2 dB for a 10-micron thick specimen at 100 MHz. That is equivalent to about 20 dB per centimeter per MHz, which is surprisingly high. And also, it's surprising that the slope of dependence on frequency is rather low. Would you like to comment on this?

SUGANUMA: Did you say that 20 dB is very large, is that right? Did you say that? It went through the specimen two times. Because it is reflected, it is to be propagated for 20 micron, in my understanding. Therefore, the attenuation coefficient per millimeter is 100 dB at 100 MHz.

HILL:...10-micron. You have 2 dB. The path length is 2 times 10-micron, so 20 micron. So that's 2 dB per 20 micron, which is 100 dB per millimeter, or 1 dB per millimeter per MHz, which is surprisingly high when the slope is not itself very high.

QUESTION: Dr. Hill, you mean our data of attenuation constant is significantly higher than the standard measurement, yes? Our data is maybe 10 times higher, and we think the difference came from the fact that we fixed the specimen with formalin, and our measurement is between 100 and 200 MHz, so the 3 MHz, and 100 MHz has maybe another kind of attenuation. We think.

THIJSSEN: May I just ask Professor Dunn to add to this discussion?

DUNN: Professor Hill, there are two arguments to this which are correct, I think. First is the way you prepare the tissue specimens can affect the attenuation. The second is that you cannot really compare the attenuation at the diagnostic frequencies with the value here, because there is a power law which is a power larger than 1.

HILL: Yes, but the power law is not showing up on the slope of the data. But let's not take a lot of time on this. I think the real question is whether your measurements do agree with independent measurements on the same specimen by a different technique.

THIJSSEN: Okay. Do you have an answer to this question, because I don't have this in mind. Is anyone able to produce a figure for 100 MHz here?

DUNN: I don't know is there's an independent measurement by different means for these specimens, but I believe they did measure other materials. I think some plastic materials were measured, and they got the same values as others did. I wanted to ask a different kind of question if I might. Can you tell us how widespread this microscope is being used for biological and medical purposes, and also for nonbiological, nonmedical purposes? Are there many of these units currently being used?

SUGANUMA: Thank you for the question. Unfortunately, this acoustic microscope is very limited in number, those which are used in medical and biological fields. I think the number currently used is less than 10 in this field in Japan.

THIJSSEN: I would like to make a question myself, if I may. You have elec-

10. Acoustic Microscope for TC

tronics, you have the receiver and the modulator, is that correct? You call it a detector? Now there are some people working with experimental acoustic microscopes, like Stuart Foster in Toronto and we have developed as well as similar kind of instrument where we used a digital oscilloscope or an instrument like that which can go up to an analog to digital conversion rate, or a traditional conversion rate of 1 GHz or so. You acquire a radiofrequency signal where you can directly measure the speed of sounds and the frequency dependence of the attenuation coefficient. Now you lose this information about the frequency dependence, so as I understood you do separate measurements for different frequencies. In the case where you could acquire the radiofrequency signal, you could get all the information for all these frequencies in one acquisition. Did you ever consider this possibility?

SUGANUMA: Thank you for the question. You talked about radiofrequency. Actually, we have used the heterodyne method for these high frequency ranges to have a conversion of frequency. Therefore, the conversion frequency range is lower than the radiofrequency range, and in terms of amplitude information, the signals are reduced to video level, so that the analog to digital converter can be used to take the information in.

THIJSSEN: Now my question is how do you get your attenuation as a function of frequency out of the signals, because I then do not understand these peculiar-shaped curves as a function of frequency. So how do you measure? You measure for separate frequencies?

SUGANUMA: The frequency can be changed in steps of 1 MHz. Therefore, by utilizing the same material we can change the frequency to obtain data. The burst wave is transmitted with a width of 500 nanoseconds, and 20 MHz bandwidth of receiver is used. Thank you for your comment.

THIJSSEN: Thank you very much again, both of the speakers.

Chapter 11

Applications of Acoustical Microscopy in Dermatology

For the last several years my laboratory at the University of California, Irvine has been exploring the biomedical applications of acoustical microscopy. One of our initial application areas was dermatology. Since this has proven to be a most useful area for the application of this technology with the potential of being utilized clinically on a regular and daily basis, I would like to share with you here some of our results and comment on the future applications of this technology for quantitatively characterizing tissue.

11.1 Equipment and Imaging Examples

The instrument used in all our studies is an Olympus UH3 Scanning Acoustical Microscope made in Japan. Figure 11.1 is a photograph of the instrument and Figure 11.2 shows a sample being placed into position on the imaging stage in our lab. By interchanging transducers, the UH3 can image at center frequencies anywhere between 30MHz and 1GHz. All of the images shown here were made at 600MHz. This frequency was selected by our pathologist for ease of comparison with corresponding optical microscope images. Figure 11.3 shows schematically the operation of the UH3 acoustical microscope. A pulse of ultrasonic energy generated by the transducer is focused through a sapphire lens and the coupling liquid onto the interface between the specimen and the stage. The amplitude of the reflected pulse is detected by the same transducer and displayed as a gray-scale boxel on the video screen. By mechanically moving the transducer over the sample, a 2-D image (a C-mode type image) is produced. In this paper, all specimens were 6 micron thick biopsy samples from human skin. The coupling liquid was water. The time required to make an image is about 5 seconds. Please

---

[*] J.P.Jones,Ph.D.

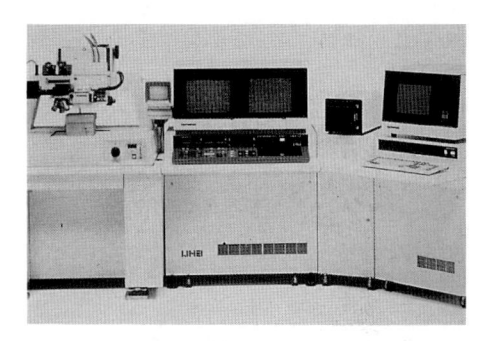

Fig. 11.1 The Olympus UH3 Scanning Acoustical Microscope used in this work

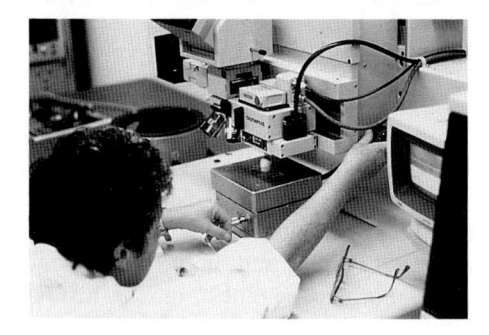

Fig. 11.2 A sample being positioned on the stage of the acoustical microscope

note that the manner in which the image is made produces a map of relative attenuation.

One of our first images of the skin and the image which greatly excited our dermatology coworkers is shown in Figure 11.4a. This figure shows a disease process known as psoriasis, which is a scaling that occurs in the skin. Figure 11.4b shows this scaling process in somewhat more detail. The dermatologists we worked with were impressed with these images because although the tissue samples were fixed, they were not stained. That is, no chemicals were applied to the specimen to bring out the scale-like features characteristic of this pathology. Such staining would have been required with optical microscopy imaging.

Figure 11.5 (a and b) shows another example of some of our early work. This is a wart-like lesion on the skin, known as seborrheic keratosis. Although this lesion is a rather nasty thing to have on your skin, the acoustical images show a beautiful flower-like structure.

To better understand the information content in the acoustical images, we have over the last few years made a detailed comparison between optical and acoustical microscope images. Figures 11.6 to 11.8 show three examples of such

11. Applications of Acoustical Microscopy in Dermatology

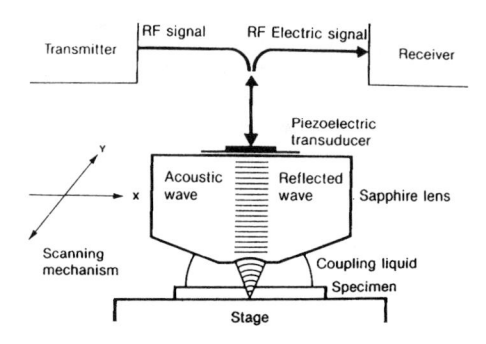

Fig. 11.3 Schematic showing the operation of the acoustical microscope

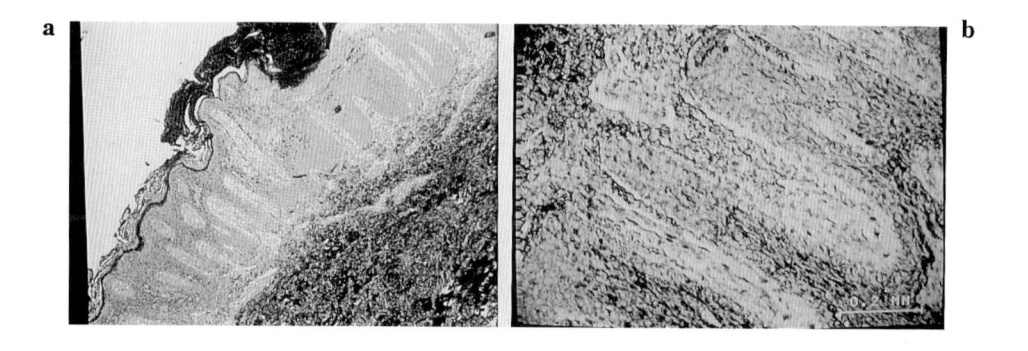

Fig. 11.4 (a) Acoustical microscope image of human skin section showing psoriasis. Note the characteristic scaling. (b) Details of scaling process (above image with smaller field of view)

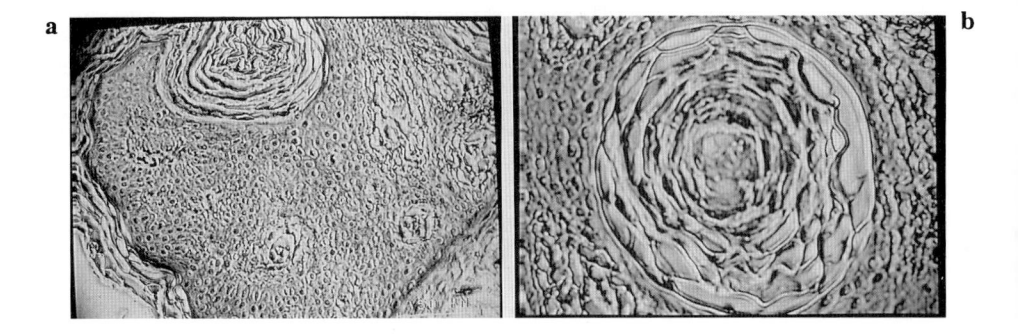

Fig. 11.5 ( )Acoustical image of seborrheic keratosis, a wart-like lesion on the skin. (b)Detail from (a)

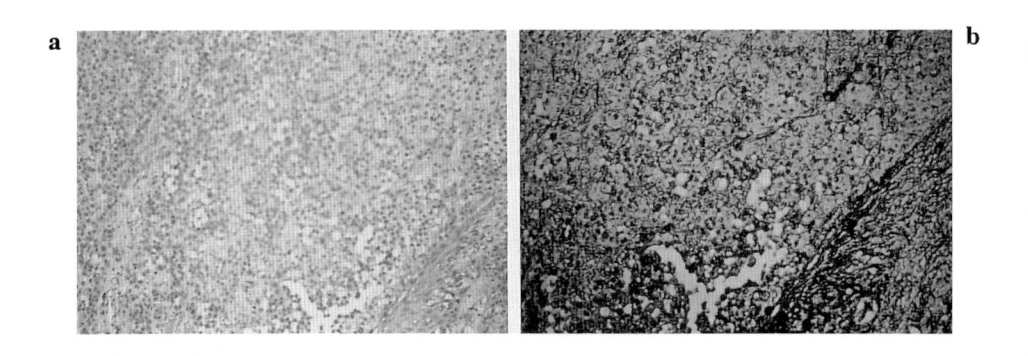

Fig. 11.6 (a)Conventional optical microscopy image of malignant melanoma, fixed and stained (H and E) section. (b)Acoustical microscopy image of adjoining section of (a), fixed but not stained. All acoustical images were made at 600MHz. Dimension across this images is about 1 mm

11. Applications of Acoustical Microscopy in Dermatology

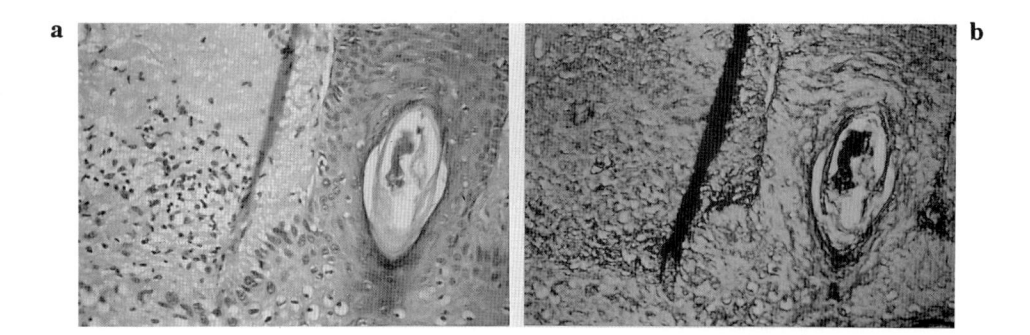

Fig. 11.7 (a)Conventional optical microscopy image of actina keratosis, fixed and stained (H and E) section. (b)Acoustical microscopy image of adjoining section of (a), fixed but not stained

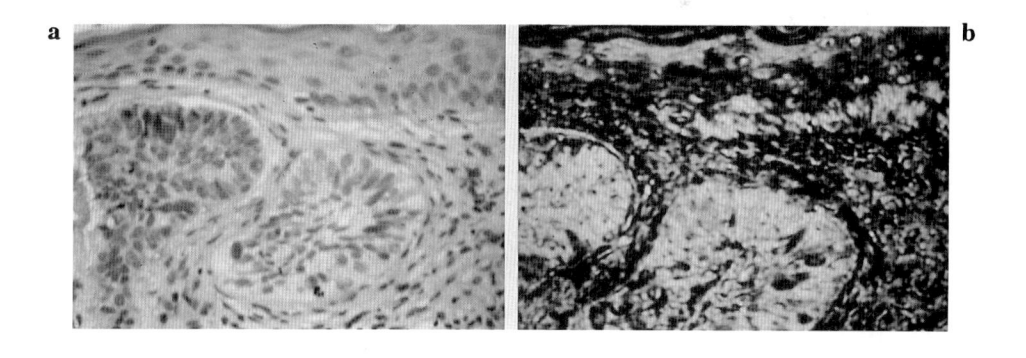

Fig. 11.8 (a)Conventional optical microscopy image of basal cell carcinoma, fixed and stained (H and E) section. (b)Acoustical microscopy image of adjoining section of (a), fixed but not stained

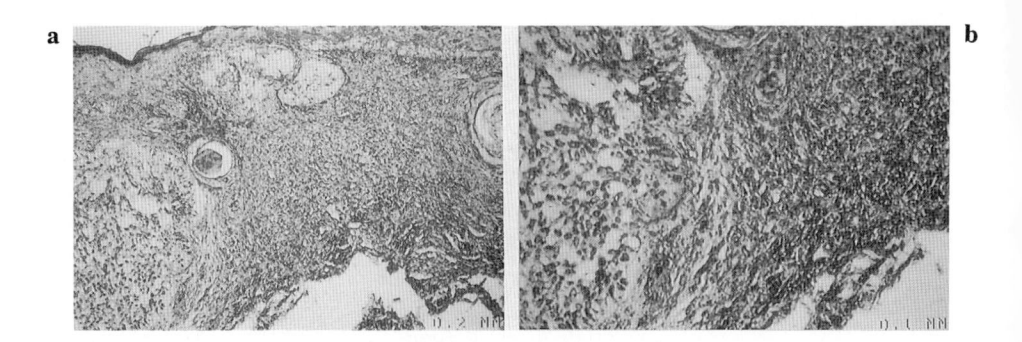

Fig. 11.9 (a) Acoustical image of a typical basal cell carcinoma (BCC). (b) Detail from (a). Note the individual basal cells

comparisons. The first (or "a") image in each of these sets is a conventional optical microscope H and E section. That is, the tissue has been fixed and stained. The second (or "b") image in these sets is an acoustical microscope image of an adjoining tissue section which was fixed but not stained. Once again, the staining process (in this case a standard hematoxylin-eosin stain) was necessary to bring out the characteristic features in the optical image. Figure 11.6 shows a malignant melanoma, Figure 11.7 actina keratosis (note the hair follicle on the right of the images), and Figure 11.8 a basal cell carcinoma. For these figures, the dimension across the bottom of each image is about 1 mm. The acoustical images (here and throughout this paper) were made at 600MHz. Although these acoustical images are quite different from their optical counterparts, they are equally diagnostic and contain sufficient information to make a pathological finding. Such has been the case for thousands of similar comparisons we have made over the years.

11.2 Acoustical Microscopy of Basal Cell Carcinoma

As an example of how acoustical microscopy can evaluate a particular pathology, the next set of figures shows a series of images of various basal cell carcinomas. Once again, all these images were made at 600MHz (although the field of view has been changed in several images to show additional detail). The tissue samples were fixed but not stained and were cut to 6 micron thickness for viewing. Figure 11.9a shows a typical basal cell carcinoma (BCC). The lower right of this figure is shown in more detail in Figure 11.9b. Note here that the individual basal cells are easily seen.

Figure 11.10a shows a keratotic BCC. The dark areas in the image represent re-

11. Applications of Acoustical Microscopy in Dermatology 207

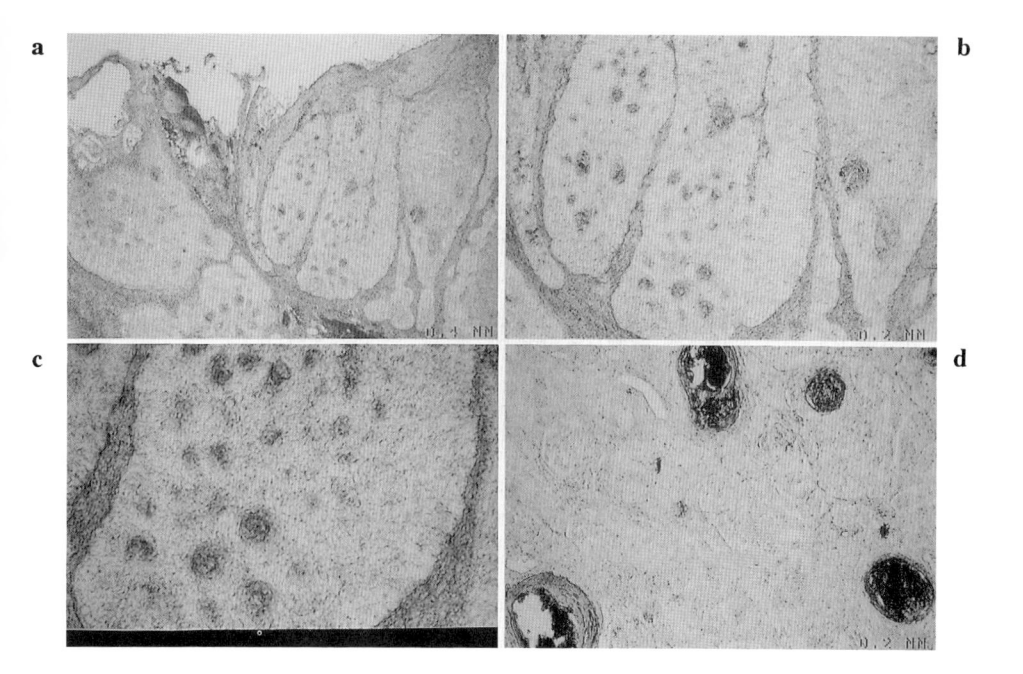

Fig. 11.10 (a)Acoustical image of a keratotic BCC. (b)Detail from (a). (c)Detail from(b). (d)Detail from (c) showing individual keratinizing foci. The white structure within the keratin centers is calcium

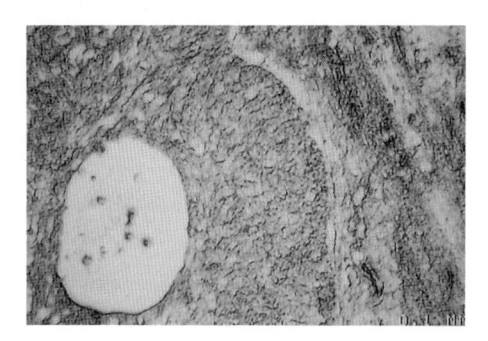

Fig. 11.11   Acoustical image of a metatypical BCC

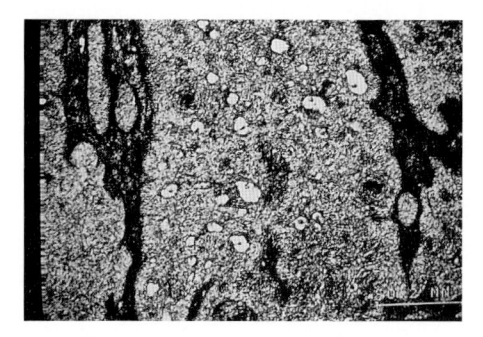

Fig. 11.12   Acoustical image of an adenoid BCC

gions of keratinizing and are better visualized here than with optical microscopy. Figures 11.10b, c and d are images of the same lesion with ever-reducing field of view. Note the individual keratinizing foci in Figure 11.10d. Here, the white structure within each foci is calcium.

A few other examples are shown in Figures 11.11 to 11.16. Specifically, 11.11 is a metatypical BCC, 11.12 is an adenoid BCC, 11.13 a nodular BCC, 11.14 a sclerosing BCC, 11.15 a pigmented BCC, and 11.16 a nodular BCC (note the central necrosis in the lesion).

Reviewing these (and many more) acoustical images of basal cell carcinoma, we can conclude that the acoustical images are distinctively different from their optical counterparts but that the acoustical images do contain sufficient information for identification of the various histological types. Keratin is particularly prominent in the acoustical images and the images are also strongly affected by the protein concentration. Thus, at least with basal cell carcinoma, acoustical microscopy is equally good as optical microscopy for identifying the various histological types. Unlike optical microscopy, acoustical microscopy allows us to

11. Applications of Acoustical Microscopy in Dermatology

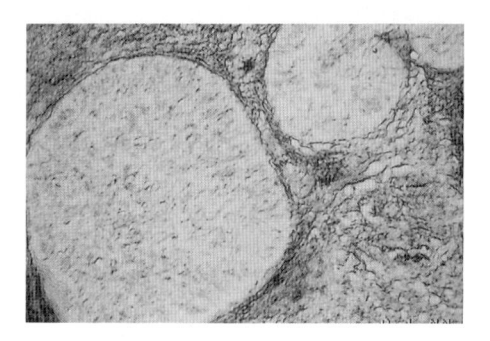

Fig. 11.13 Acoustical image of a nodular BCC

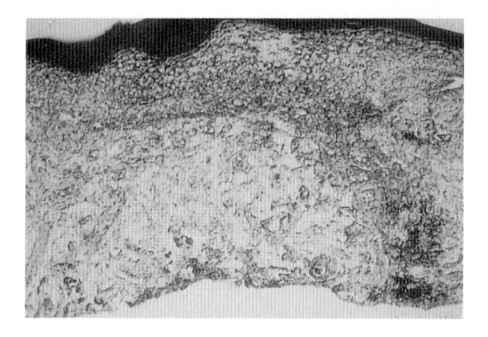

Fig. 11.14 Acoustical image of a sclerosing BCC

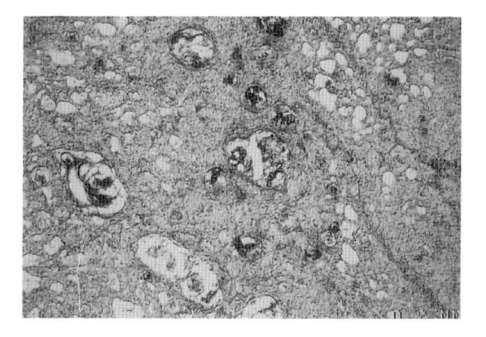

Fig. 11.15 Acoustical image of a pigmented BCC

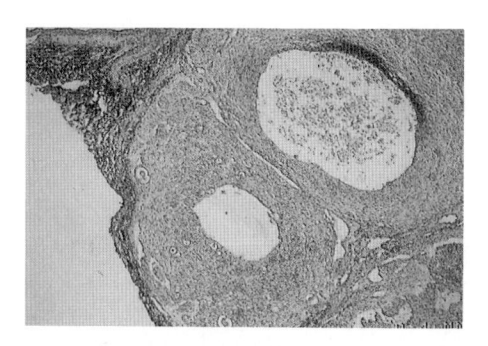

Fig. 11.16 Acoustical image of a nodular BCC

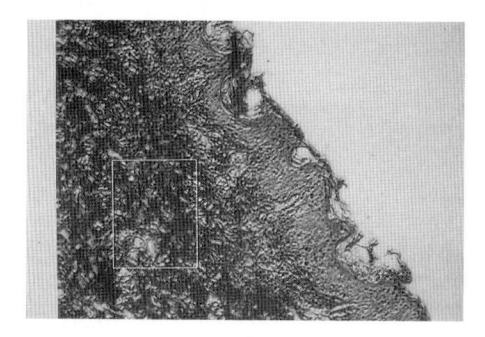

Fig. 11.17 Acoustical image of sun-damaged skin. Computer-selected window allows computation of mean echo amplitude, which is a measure of the relative attenuation in the region

make this identification without tissue staining or elaborate sample preparation.

11.3 Quantitative Methods and Conclusions

In the discussion related to Figure 11.3, it was pointed out that the images obtained with the UH3 acoustical microscope are essentially mappings of attenuation. Thus, by selecting a region of interest and computing the mean echo amplitude in that region, we obtain a measure of the average attenuation in that region. Figure 11.17 shows an acoustical microscope image of a biopsy specimen of sun-damaged skin. Note the computer-generated region of interest in which we can easily measure the mean echo amplitude (or, equivalently, the mean attenuation). Applying this measure to a number of normal and sun-damaged skin samples we obtain the results given in Figure 11.18. Thus, the effect of the sun on both the dermis and epidermis can be quantitatively described in terms of

11. Applications of Acoustical Microscopy in Dermatology

Skin Type	Relative Attenuation
dermis: normal (10)	35 ± 2.1
dermis: "some" sun damage (5)	41 ± 3.3
dermis: sun damaged (22)	46 ± 2±8
epidermis: normal (10)	15 ± 1.3
epidermis: "some" sun damage (5)	27 ± 3.2
epidermis: sun damaged (22)	34 ± 2.5

Fig. 11.18 Relative attenuation values for normal and sun-damaged skin

Fig. 11.19 A cartoonist's view of early optical microscopy

Fig. 11.20   Our view of early acoustical microscopy

attenuation. This same method can be used to study the effects of drugs and / or chemicals on the skin. Since certain models of the skin can be grown in culture, acoustical microscopy provides a means to conduct laboratory tests on the effects of drugs and various environmental factors.

Figure 11.19 is a cartoonist's view of early optical microscopy. I would suggest that if this is true, then Figure 11.20 represents early acoustical microscopy. We have only begun to explore the biomedical applications of this technology. Our initial work suggest that this technology has a great future not only in biology but in many areas of diagnostic medicine.

Discussion

DUNN: Early on you said that you were using neighboring sections. You sectioned material for the optical microscope, and then you used neighboring ones for the acoustic microscope. So those were fixed, I take it, in something like formalin? Were they also embedded in paraffin? Or was the paraffin removed for the acoustic microscopy imaging?

JONES: That's a good question. We have done some work with paraffin sections, and those have been very difficult to use, because it removes some of the material. Generally, you want to de-paraffinize before you look with the acoustical microscope. That actually removes a lot of material you'd like to see. These sections on the images I showed you here were all fixed and frozen, but they were not a paraffin block. These were essentially frozen as a block of tissue, and then sliced with a microtone, which is not a standard preparation.

DUNN: But they were fixed?

JONES: Yes, they were fixed. We've looked at a number of samples that were not fixed, just frozen. And the images are actually much better. We would prefer to work with those, but the dermatologists aren't terribly willing to do that. They much prefer, because of the handling of the tissue, to have fixed tissue. We have in a number of cases gotten fresh tissue to work with, particularly from animal specimens, that's a much better way to look at tissue with the acoustical microscope.

DUNN: Early in your talk you were stressing the fact that the acoustic images did not have to be stained. Later on, I gather that you weren't so strong about that; that the staining wasn't really the important thing. There has to be additional information available in the acoustic microscope, and it's not simply the step of staining that makes it valuable?

JONES: I think what makes it valuable is not just staining. We have been looking for ways essentially to sell this to the pathologists, to get them interested in looking at acoustical images. Frankly, we have had some difficulty because many pathologists claim that the optical equipment is sufficient for their needs and why should they look at these other slightly funny images? One of the selling points early on was that, from some of the initial studies, they were able to see very interesting features without staining. Depending on the pathology, there can be even a day or so delay in getting a diagnosis. What has worked at our institution is that we have an arrangement with the out-patient facility for dermatology, in the next building from our lab, and they essentially just bring over these sections and look at them with the microscope and make a diagnosis right there. It's very convenient for them.

DUNN: An advantage of optical microscopy is the fact that one does see cellular structure. I couldn't tell whether you saw cellular structure with the acoustic imaging figures that you showed.

JONES: Not at these frequencies. At higher frequencies you can begin to see some cellular structure, but in this particular example and dealing with this particular group of pathologists, this seemed to be the kind of images they would like to look at. Other work that we have recently done with a cardiac pathologist looking at biopsy specimens from myocardium at higher frequencies, you can see cellular structures, and that does appear to be of some interest.

HILL: On the matter of staining, has anyone thought of developing specific acoustic stains? Because when you fix with formalin, for example, you are in effect staining the tissue. You are altering the protein, in particular, in a manner that changes the acoustic contrast.

JONES: That's right.

HILL: I hadn't really thought of this, but there may be other processes that you could use that would bring out other features of the tissue.

JONES: That's a very interesting comment. No, actually, we haven't thought about acoustical stains, but I think it would be something very interesting to do. There's no question that staining does alter the features of tissues. Our observations are more anecdotal than serious study of comparison. But certainly, when you look at tissue specimens that have been fixed and similar tissue specimens that have not been fixed, the tissue specimens that have not been subjected to the fixative, yield much better images. You seem to see a lot more detail.

TANAHASHI: I'm Dr. Tanahashi from Tohoku Hospital. I thank you for your nice presentation. I'd like to ask some questions. I'm a surgeon. For us surgeons, the technique you have introduced today, namely, acoustic microscopy, is quite interesting. I look forward to its clinical application as soon as possible. Between benign lesions and malignant lesions, if you use HE staining, we can look at differentiation. But compared to differentiation diagnosis by HE staining, how does this technology compare–is it better or worse? Which is better–HE or this one, in differentiating benign lesions from malignant lesions? The second question. During the operation, we take a small amount of tissue and we want to look at the tissue to see whether it's malignant or benign. We want to decide during the operation what to do– whether to excise extensively or moderately. And that is going to be helped by this kind of technique, I hope. If that is the case, for the second case, intersurgical diagnosis, you have looked at the frozen section as well, right? But after the process of freezing, there's damage to the tissue. In other words, for the frozen section diagnosis during the surgery, the diagnostic value is inferior compared to HE stain diagnosis, because of the damage caused by the freezing process. You've looked at both frozen sections and fixed sections, was there a difference, or were they equally good?

JONES: Those are several very good questions. Let me see if I can remember them. The first one dealt with the comparison of the acoustical images with conventional H and E stained sections, if I recall correctly. And whether one could make a judgment whether the acoustical images are better, or the H and E sections are better. I couldn't really say. The only comment that I could make is that our pathologists find that the acoustical images, which once again are made with fixed but not stained tissue, are equally diagnostic to those of a conventional H and E sections. I don't think one could make a comment, at least at this time, about which would be better. There are a number of us, who would feel that there is probably, just based on simple physics, there is probably a lot more information in the acoustical images. But it's going to take a number of years before that's fully appreciated, or before that kind of information content is identified. So at the present time, all we've done, basically is make a comparison between the H and E sections and the acoustical images, and all that we can say is that the acoustical images are equally diagnostic. You made some comments about the effects that freezing has on tissue, and you're absolutely correct. Freezing the tissue does alter its structure. And we would certainly prefer to deal with fresh tissue that is not stained. We have only a few examples of cases where we have been able to look at fresh tissue. The problem is preparing a fresh tissue specimen of a few microns thick without freezing it or preparing it in some way. If you try to slice fresh tissue with that thinness, it's practically impossible. So perhaps some other techniques could be developed that would produce thin sections to look at with the microscope. I think that in the future, one would hope that various types of acoustical imaging devices could be developed working at relatively high frequencies of several hundred megahertz that you could use during surgery so that you could place the transducer directly on the tissue and produce images practically as good as the ones I showed here. That certainly would seem to be in the cards from a technological standpoint. It's one that I think a number of groups are at least thinking very seriously about preparing instruments to work in that context and that kind of frequency range. So the real potential of acoustical microscopy is not only as a laboratory technique and as a clinical tool for looking at specimens, but also as a possibility of using that technology in situ, in the operating room directly on the patient, to produce images of sufficient quality that would obviate the need for either doing a biopsy or perhaps even exploratory surgery.

TANAHASHI: Thank you very much for answering my questions. I hope that there will be more progress in this technology due to your very industrious efforts. Thank you very much.

THIJSSEN: This morning, we had a similar discussion. Is quantification necessary? I believe it is very important for tissue characterization. For optical microscopy, with the technique of staining, in other words by using histochemical reactions, we are looking at the possibility of quantification by using a bulk chemical approach. Regarding this, with acoustical microscopy we can look at the acoustic characteristics. In other words, the biological tissue reaction to the elastic wave, right? In that regard, I believe that physical characteristics can be picked up by acoustic microscopy, which cannot be picked up by optical methods. But if we want to quantify that, then we have to think about how to do that, as Dr. Hill pointed out this morning. First of all, if stain/fixing changes acoustic characteristics, then we have to try and use some new technique in order to improve the enhancement technique. For that purpose, we use different fixatives, like formalin and the others, in order to see how the tissue will be altered. By using different fixatives, we are not yet ready for quantification. But I believe that kind of enhancement technique is necessary for acoustic microscopy, as well. For the sake of the quantification, what approaches are you thinking about for quantification using acoustic microscopy? You have shown the relative attenuation data. But what is the reference you use in order to make a judgment on the relative attenuation?

JONES: OK. Well, there is a question about what you mean by quantification. I think there are degrees of quantification that one can have. I guess there is some point as to what we really mean by ultrasonic tissue characterization, as Professor Hill indicated this morning. I think there are a range of things and a range of degrees of quantification that one can have. One can make simple B-mode images that are very qualitative images. With the acoustical microscope, although we are making images, they are basically images of attenuation. We're not making a quantitative measure of attenuation, but we're certainly making images that are closely related to attenuation. Relative attenuation values may be sufficient for looking at certain pathologies. To get absolute values of attenuation may be very difficult with the acoustic microscope. One could presumably make velocity images with something like the acoustic microscope. There is probably a great deal of information in the kind of texture patterns that you see in the pathologies, particularly as Dr. Thijssen indicated today. Some of those techniques are applicable to acoustical microscopy. Thank you.

Chapter 12

High Frequency Acoustic Properties of Tumor Tissue

12.1 Introduction

Ultrasonic tissue characterization of tumor tissue is important for two reasons; first that the acoustic parameters provide the physical mechanical properties not determinable by for example, optical microscopy, and second that such data provides information for understanding echographic imaging.

In the present study, a scanning acoustic microscope (SAM) system was equipped to measure the acoustic properties of tumor tissue at the microscopic level.

12.2 Materials and Method

The tumor tissues involved in this study were gastric cancer and renal cell carcinoma. All tissues were surgically excised, formalin-fixed, paraffin-embedded, sectioned $10\mu m$ in thickness, and mounted on glass slides for SAM measurements. The paraffin was removed by the graded alcohol method just before the ultrasonic measurement.

A specially developed scanning acoustic microscope system, operating in the frequency range of 100 to 200 MHz, was employed, and the images of the two dimensional distribution of attenuation constant and sound speed were displayed.

Figure 12.1 is a block diagram of the SAM system. The transducer is oscillated along the x–axis, and the sample holder is scanned along the y–axis. The distilled water coupling medium was maintained at 20 ℃, as was the tissue specimen, during the measurement procedure.

The analogue image processor stores the eleven frames of amplitude images in 10 MHz steps in the range 100 to 200 MHz, and five phase images in 10 MHz

---

[*]Yoshifumi Saijo, M.D., Ph.D. and Hidehiko Sasaki, M.D.

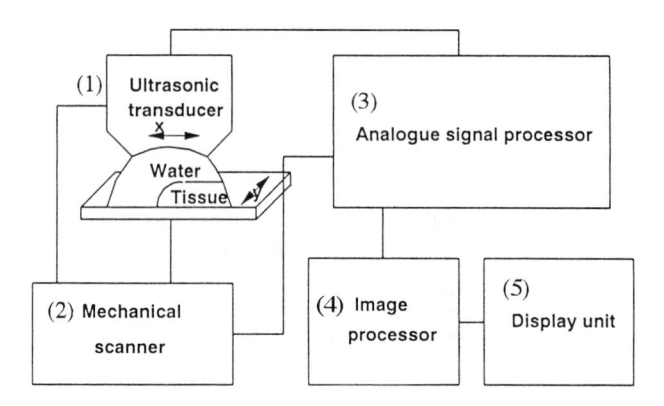

Fig. 12.1 Block diagram of the scanning acoustic microscope system. The system consists of five parts, viz., (1)Ultrasonic transducers, (2)Mechanical scanner, (3)Analogue signal processor, (4)Image signal processor, and (5)Display unit

steps in the range 100 to 140 MHz. The thickness of the specimen is determined from the frequency dependent characteristics of the amplitude and the phase of the received signals of the sixteen frames [1].

The values of attenuation constant and sound speed were obtained after the thickness was determined. Table 1 shows the relationship between color coded scales and values of attenuation constant and sound speed.

Two dimensional distributions of attenuation constant and sound speed are displayed on the CRT monitor. 480×480 pixels comprise one frame in this system.

The region of interest (ROI) was determined by comparison of optical and acoustic images. The size of the ROI was $200 \times 200 \mu m$ in the optical microscopy image and 50×50 sampling points in the acoustic images.

12.3 Results - Gastric Cancer

Gastric cancer is very prevalent in Japan; the incidence being is about 1000 cases per million of population. The five year survival rate is about 95 % in the case of early stage gastric cancer detection. It is thus very important to detect cancer in an early stage and to distinguish malignant cells from benign tissues.

Endoscopic ultrasonography, which is undergoing development, has observed submucosal invasions of cancerous tissue not detected by standard endoscopic biopsy. However, endoscopic ultrasonography cannot be used as yet for the classification of gastric cancers. To determine the acoustic properties of each type of the gastric cancer tissue is thus important for understanding the clinical appearances of gastric cancer.

The gastric cancer tissues were classified into five groups according to their

12. High Frequency Acoustic Properties of Tumor Tissue

TUBULAR ADENOCARCINOMA
(WELL-DIFFERENTIATED)

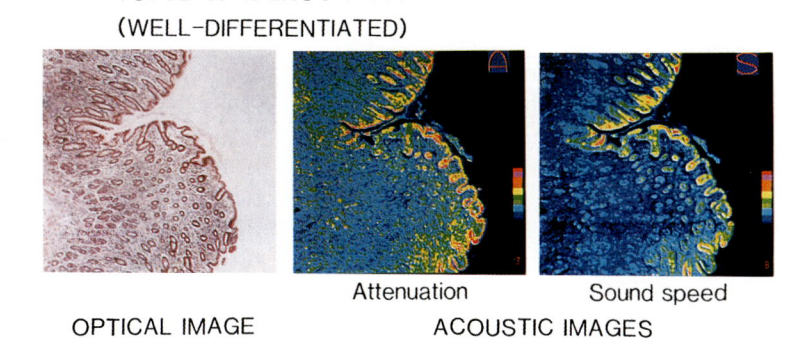

OPTICAL IMAGE   Attenuation   Sound speed
ACOUSTIC IMAGES

Fig. 12.2 Optical (left, H-E stain, ×40) and acoustic (center: attenuation constant, right: sound speed) images of a specimen of well-differentiated tubular adenocarcinoma. The color bar in each acoustic image represents quantitative values of the attenuation constant and sound speed in Table 12.1

pathological findings, viz., papillary adenocarcinoma, well-differentiated tubular adenocarcinoma, moderately differentiated tubular adenocarcinoma, poorly differentiated adenocarcinoma, and signet-ring cell carcinoma.

Four specimens of each type of the cancer were investigated; thus 20 specimens from 20 patients were involved in this study [2].

Figure 12.2 shows the optical and acoustical images of a specimen of well-differentiated tubular adenocarcinoma. The color bar scale of each acoustic image relates to values of attenuation constant and sound speed. The figure shows that the attenuation constant varies from 1.0 to 2.25 dB/mm/MHz and that the sound speed is in the 1600 to 1750 m/s range in the cancerous lesion.

Figure 12.3 shows a case of moderately differentiated tubular adenocarcinoma.

Figure 12.4 is a case of poorly differentiated tubular adenocarcinoma. The structure of the tumor is poorly organized, and the both acoustic parameters indicate lower values than those of well-differentiated tubular adenocarcinoma.

Figure 12.5 is a case of papillary adenocarcinoma. Both attenuation and sound speed are nearly the same as those for the normal mucosa.

Figure 12.6 shows a case of signet-ring cell carcinoma. The infiltration of signet-ring cells are seen in the submucosal layer. Both the attenuation constant and the sound speed are low in this type of gastric cancer tissue.

Figure 12.7 shows the means and the standard deviations of all the specimens investigated. For the signet-ring cell carcinoma specimens, the mean attenuation constant is 0.49 dB/mm/MHz and the sound speed is 1523.3 m/s, which are significantly lower than those of the normal mucosa and of the other types of gastric cancer tissues, except for poorly-differentiated adenocarcinoma.

Attenuation constant (dB/mm/MHz)	Color	Sound speed (m/s)
1.9 ~	Red	1690 ~
1.7 ~ 1.9	Magenta	1670 ~ 1690
1.5 ~ 1.7	Orange	1650 ~ 1670
1.3 ~ 1.5	Brown	1630 ~ 1650
1.1 ~ 1.3	Yellow	1610 ~ 1630
0.9 ~ 1.1	Green	1590 ~ 1610
0.7 ~ 0.9	Olive green	1570 ~ 1590
0.5 ~ 0.7	Cyan	1550 ~ 1570
0.3 ~ 0.5	Royal blue	1530 ~ 1550
0.1 ~ 0.3	Blue	1510 ~ 1530
~ 0.1	Black	~ 1510

Table 12.1 Relationship between color coded scales and values of attenuation constant and sound speed

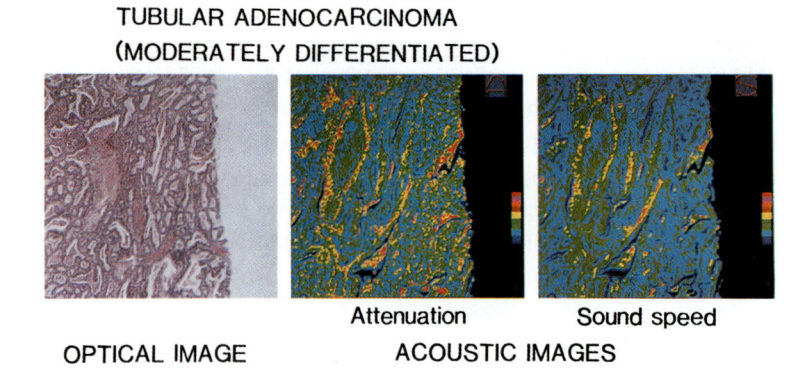

Fig. 12.3 Optical and acoustic images of a specimen of moderately differentiated tubular adenocarcinoma

12. High Frequency Acoustic Properties of Tumor Tissue 221

POORLY DIFFERENTIATED ADENOCARCINOMA

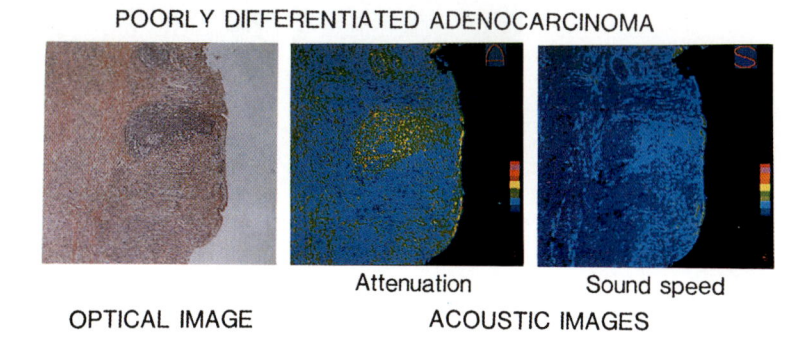

OPTICAL IMAGE   Attenuation   Sound speed
                ACOUSTIC IMAGES

Fig. 12.4 Optical and acoustic images of a specimen of poorly differentiated adenocarcinoma

PAPILLARY ADENOCARCINOMA

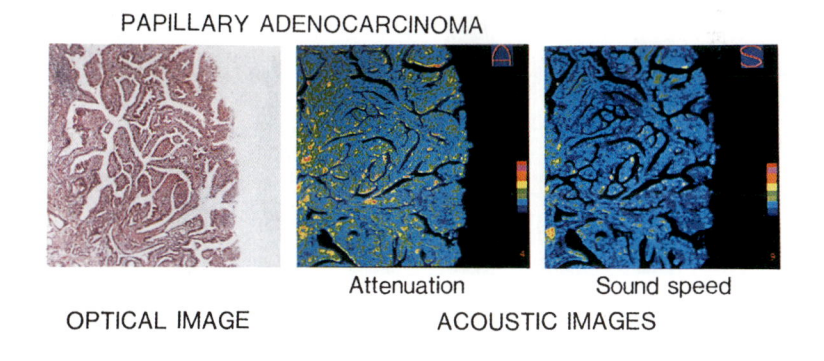

OPTICAL IMAGE   Attenuation   Sound speed
                ACOUSTIC IMAGES

Fig. 12.5 Optical and acoustic images of a specimen of papillary adenocarcinoma

SIGNET-RING CELL CARCINOMA

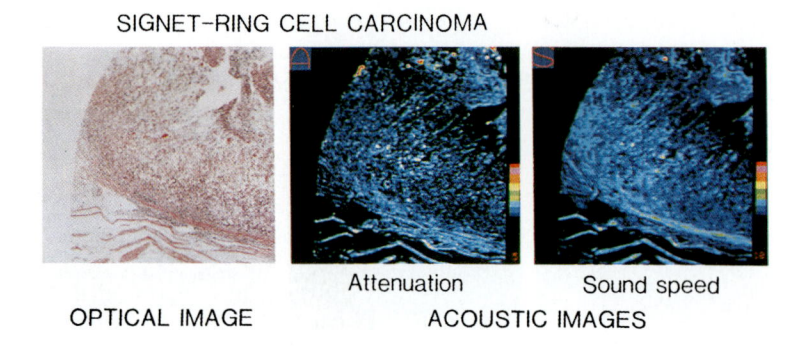

Fig. 12.6 Optical and acoustic images of a specimen of signet-ring cell carcinoma. The infiltration of the signet-ring cells are observed in the submucosal layer

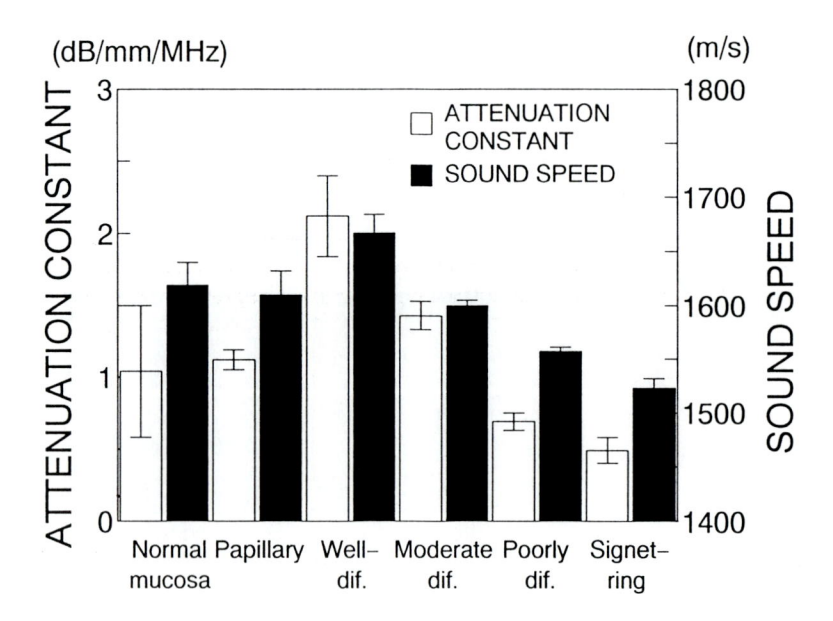

Fig. 12.7 Graph showing the means and the standard deviations of all the gastric cancer specimens. Papillary: papillary adenocarcinoma, Well-dif.: well differentiated tubular adenocarcinoma, Moderate dif.: moderately differentiated tubular adenocarcinoma, Poorly dif.: poorly differentiated adenocarcinoma, Signet-ring: signet-ring cell carcinoma

12.4 Discussion - Gastric Cancer

The values of the attenuation constant and the sound speed increased as the cellular differentiation proceeded through the three kinds of tubular adenocarcinoma. As the density of the biological soft tissues can be assumed to be nearly constant, increased sound speed can thus be interpreted to mean that tubular adenocarcinoma tissues become acoustically stiffer as the differentiation of the tissue proceeds. Electron microscopy has shown that the number of desmosomes, which are considered to attach cell-to-cell, is significantly decreased in poorly differentiated adenocarcinoma. Well-differentiated tubular adenocarcinoma specimens exhibit nearly the same number of desmosomes as in normal mucosal tissue. This increasing trend was thus regarded as the result of tightening of the intercellular attachment [3, 4].

Both the attenuation constant and the sound speed were significantly lower in the signet-ring cell carcinoma than in the adenocarcinoma. The intracellular component of the signet-ring cell carcinoma is the periodic acid, Schiff stain (PAS) positive substrate. The lower values of the attenuation and sound speed may be accounted for by the intracellular chemical components of the tumor tissues.

These data of gastric cancer tumors clearly show that the SAM system can be used to classify the types of cancer tissues, as revealed by measurement of the acoustic parameters of the pathologies.

12.5 Results - Renal Cell Carcinoma

Renal cell carcinoma is known to exhibit both low and high intensity echoes in clinical echography [5]. Although the relationship between the echographic appearance and the histological appearance has been studied previously, the earlier studies did not reveal the nature of this relationship clearly [6]. Thus, measurements of the acoustic properties at the microscopic level were considered to be important for understanding the clinically obtained echographic features.

Renal cell carcinoma is classified into two groups; clear cell subtype and granular (dark) cell subtype [7].

Figure 12.8 shows an example of clear cell subtype of renal cell carcinoma. The values of the attenuation constant and sound speed are, respectively, 0.4 dB/m/MHz and 1520 m/s. These values are lower than those of normal renal cortex, as shown in Figure 12.9.

Figure 12.10 shows a case of granular cell subtype of renal cell carcinoma. Both acoustic parameters are also lower than those of normal renal cortex, viz., 0.4 dB/mm/MHz and 1540 m/s.

Figure 12.9 shows the means and the standard deviations of all the renal data [8]. The values are 1.15 dB/mm/MHz and 1624 m/s for normal renal cortex. Both parameters have significantly lower values for clear cell and granular cell subtypes than those of normal cortex. No significant difference is seen between

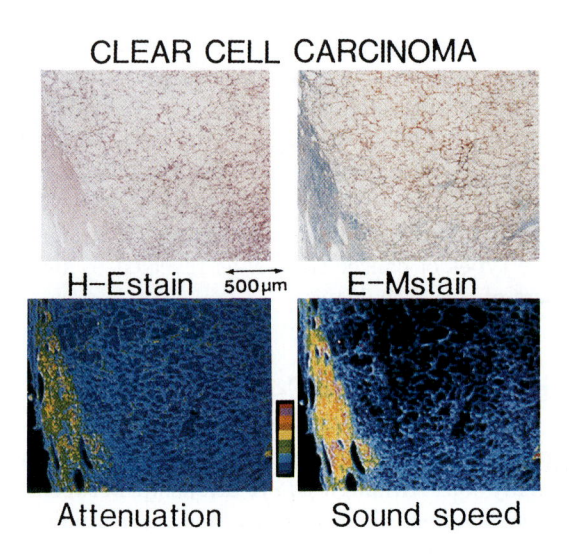

Fig. 12.8 Optical and acoustic images of a specimen of clear cell subtype of renal cell carcinoma

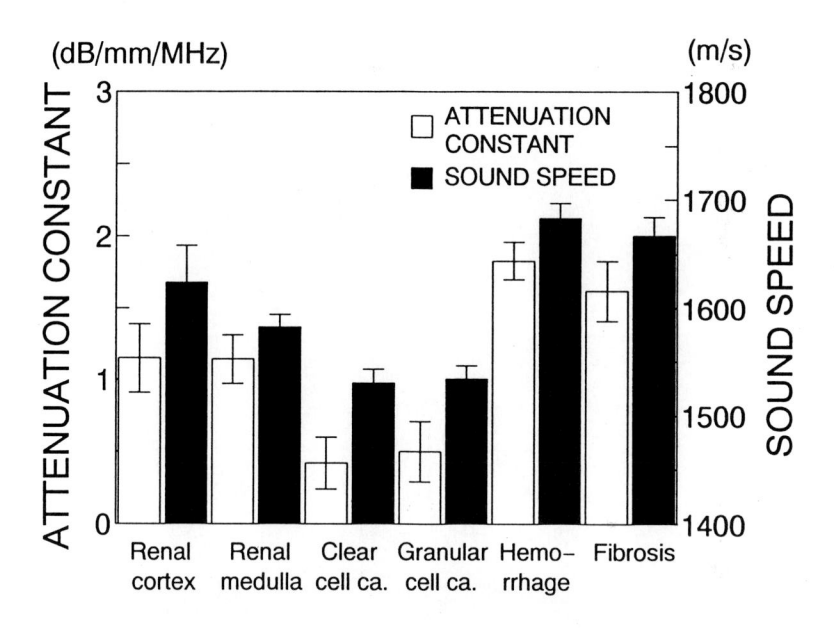

Fig. 12.9 Graph showing the means and the standard deviations of all the renal cell carcinoma specimens. Clear cell ca.: clear cell subtype of renal cell carcinoma, Granular cell ca.: granular cell subtype of renal cell carcinoma

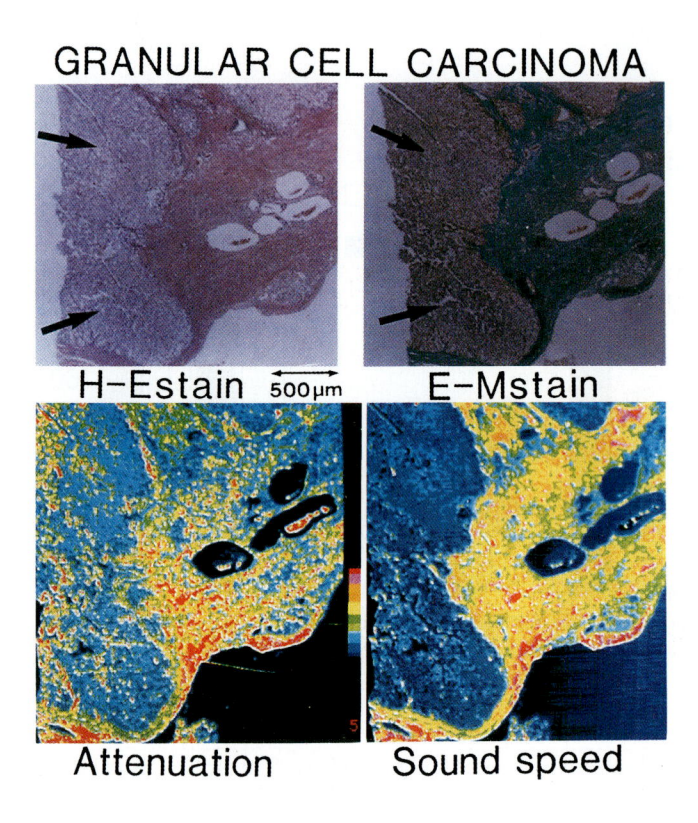

Fig. 12.10 Optical and acoustic images of a specimen of granular cell subtype of renal cell carcinoma

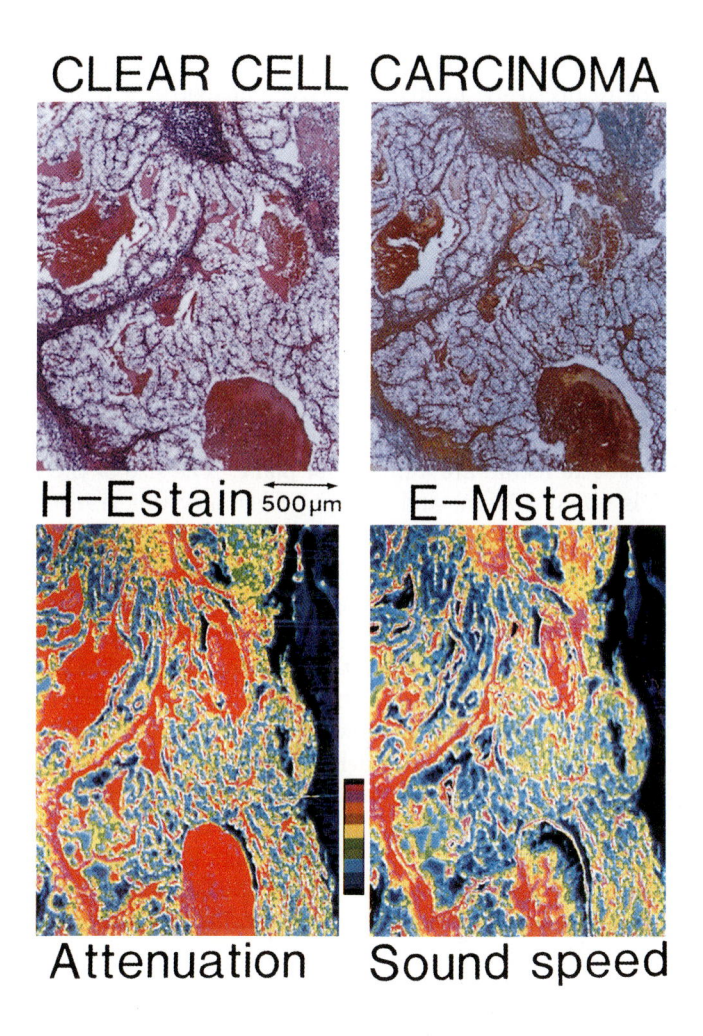

12. High Frequency Acoustic Properties of Tumor Tissue

Fig. 12.11 Optical and acoustic images of a specimen which exhibited very strong echo in the clinical echography. Hemorrhagic lesions represents high values of the attenuation constant and sound speed

these two subtypes of renal cell carcinoma.

Both the attenuation constant and the sound speed show significantly higher values for hemorrhagic and fibrotic tissues.

Figure 12.11 is a case which shows very strong echoes in the renal tumor in clinical echography. Optical microscopy shows that the tumor is comprised of clear cell, hemorrhage, and fibrosis. The values of the attenuation constant and sound speed are, respectively, 1.8 dB/mm/MHz and 1690 m/s in the hemorrhagic lesion, while both values are very low in the clear cell subtype.

12.6 Discussion - Renal Cell Carcinoma

The interface echo separating a pair of media is explained in terms of the magnitude difference of the two specific acoustic impedances. The difference of the acoustic impedance between clear cell and hemorrhage is very large, thus, a strong echo may be produced at the interface between the two types of tissue elements. However, the ideal reflection is usually discussed with the understanding that the fluid-like media are infinite and plane. The size of the hemorrhage in Fig. 12.11 is about 500μm, and it is not clear how this size influences the reflection.

The intratumorous strong echo is also not explained by the interface echo. Thus, another interpretation for the strong echo is required involving the relation between the acoustic properties and the histological structures. The acoustical images in Fig. 12.11 show the apparent island-like distribution of the hemorrhagic lesion, exhibiting high values of the attenuation and sound speed, in the base of the clear cell tissues exhibiting low values of the both acoustic properties. The acoustic field in the tumor is considered to be inhomogeneous for the 3 MHz ultrasound of clinical investigation, because the size of the hemorrhagic lesion and the wave length are nearly same in this case. Our data suggest that the strong echo in the tumor may be related to the inhomogeneity of the acoustic field in the tumor, consisting of the clear cell and the hemorrhage, and that some kinds of scattering may be increased in such a medium to lead to strong echoes in clinical echography.

12.7 Conclusions

The acoustic properties of renal and gastric cancer tissues were obtained with a scanning acoustic microscope system. The data suggest that the measurement of acoustic properties at the microscopic level is useful to characterize the types of gastric cancer by the physical properties reflected in the acoustic parameters, and that the clinical echographic appearances of the renal tumor is related to the inhomogeneous acoustic field in the tumor.

References

1. Okawai H, Tanaka M, Dunn F, Chubachi N, Honda K (1988) Qualitative display of acoustic properties of the biological tissue elements. Acoustical Imaging 17: 193-201

2. Saijo Y, Tanaka M, Okawai H, Dunn F (1991) The ultrasonic properties of gastric cancer tissues obtained with a scanning acoustic microscope system. Ultrasound Med and Biol 17: 709-714

3. Goldman H, Ming SC (1968) Fine structure of intestinal metaplasia and adenocarcinoma of the human stomach. Lab Invest 18: 203-210

4. Nevalainen TJ, Jarvi OH (1977) Ultrastructure of intestinal and diffuse type gastric carcinoma. J Path 122: 129-136

5. Fujii H, Kaneko S, Hashimoto H, Sasaki M, Yachiku S (1989) Ultrasonotomograms of renal cell carcinoma -Retrospective study of the internal echogram of tumor. Jpn J Med Ultrasonics 16: 375-382 (in Japanese)

6. Jinzaki M, Hisa N, Fujikura Y, Ohkuma K, Tashiro Y, Sugiura H (1990) Comparative study between ultrasonographic and pathohistological findings of small renal cell carcinoma, Jpn J Med Ultrasonics 17: 280-287 (in Japanese)

7. Bennington JL, Beckwith JB (1975) Atlas of tumor pathology. Tumors of the kidney renal pelvis and ureter. AFIP, Washington DC

8. Sasaki H, Saijo Y, Naganuma T, Tanaka M, Terasawa Y (1993) Acoustic properties of renal cell carcinoma. Jpn J Med Ultrasonics 20 suppl.2: 345-346 (in Japanese)

Chapter 13

Acoustic Properties of the Fibrous Tissue in Myocardium and Detectability of the Fibrous Tissue by the Echo Method

13.1 Introduction

Some heart diseases, such as cardiomyopathy, myocardial infarction, and others, are accompanied by myocardial damage. When advanced diagnostic equipment is employed in the clinical examination of a patient with myocardial damage, intensified abnormal echoes, at the damaged myocardial area, are frequently observed. Figure 13.1 shows two dimensional echocardiograms taken from a case with hypertrophic cardiomyopathy(HCM). Abnormal strong speckle-like echoes are seen in the area of the left ventricle. However, the mechanisms responsible for the occurrence of these echoes remain obscure.[1]

This paper reports investigations performed to determine whether or not the boundaries of the abnormal tissues, appearing in the myocardium, are the echo source of the speckle-like echoes. The data were obtained by the scanning acoustic microscopy method (SAM) developed in our laboratory.

---

*M. Tanaka,M.D. and Floyd Dunn,Ph.D.

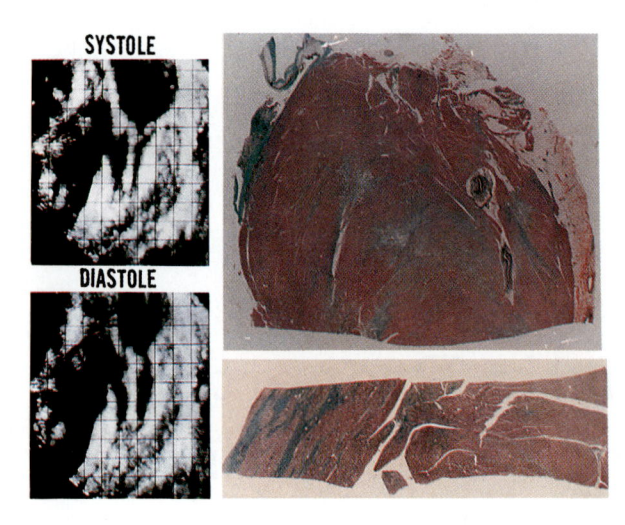

Fig. 13.1 Two dimensional echo cardiograms and optical microscopic images of the myocardium at corresponding areas of the left ventricular wall. The specimens were taken at autopsy in a case of hypertrophic cardiomyopathy

13.2 Method and Materials

When the width of the ultrasonic beam is greater, and the wave length is longer, than the significant structural features of the tissue elements, the resulting acoustic characteristics of the tissue exhibit a gross or spatial average of numerous individual elements within the ultrasonic beam width of the normal and abnormal tissues or of various kinds of tissue elements, such as muscle fiber, connective tissue, vessel tissue, etc. Details of the acoustic characteristics of normal tissue structural components, and those of abnormal tissues, can be obtained by employing the high frequency narrow beam ultrasound of the SAM methods.[2)3)]

Figure 13.2 shows a block diagram of the SAM - type acoustic microscope. The diameter of the ZnO transducer is 1.6 mm, and the ultrasonic beam converges conically through a concave sapphire lens with a 1.25 mm radius of curvature. The frequency range of the pulsatile ultrasound can be chosen from 100 to 200 MHz. The resolution of the microscope is of the order of 8 to 10 micron.

The formalin fixed, or frozen, specimen of about 10 microns in thickness is placed on the glass plate at the focal plane of the transducer. A small quantity of water is used as the acoustic coupling medium.

The rate of attenuation and the speed of ultrasound in the tissue specimens are measured from the two dimensional image obtained by the high speed mechanical scan in the two directions of X (at a speed of 125 to 250 mm/s) and Y (at a speed of 0.5 mm/s) normal to the beam; the so called C-scan. The rate of attenuation in the specimen is calculated from the amplitude of the reflected signal from the posterior surface of the specimen in comparison with the amplitude of the

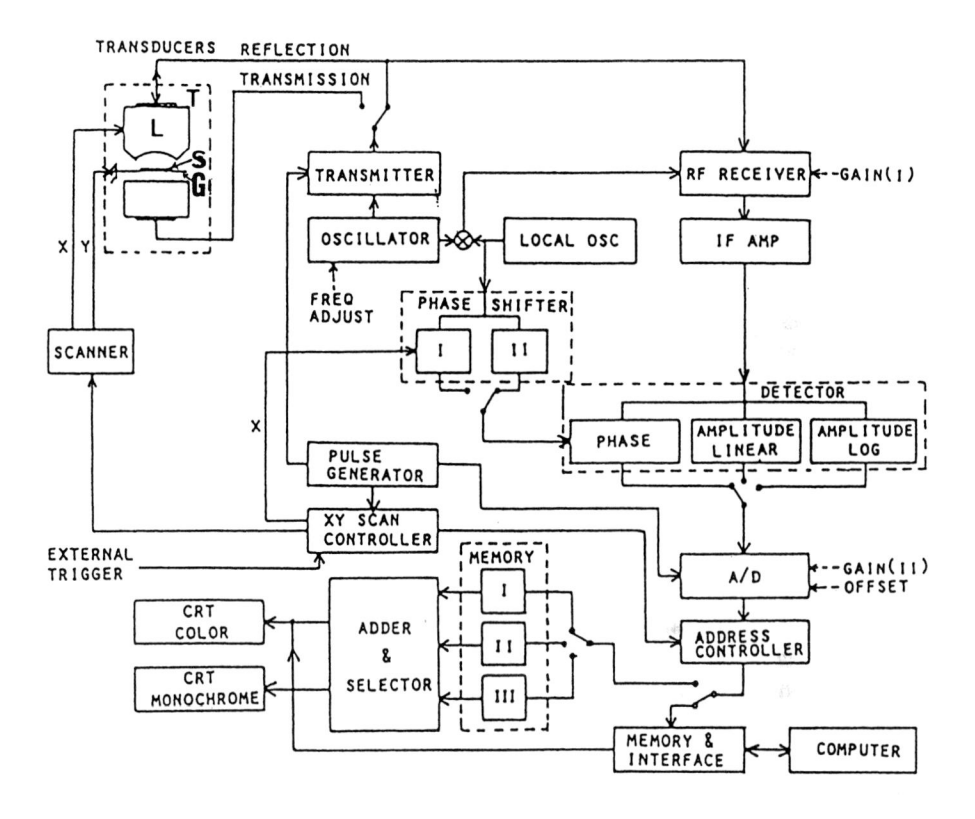

Fig. 13.2 Block diagram of the Scanning Acoustic Microscope (SAM) developed in our laboratory. T : transducer, L : sapphire lens S : specimen, G : glass plate

Color	Attenuation(dB/mm/MHz)	Velocity(m/s)
red	≥ 22.375	≥ 1765
magenta	2.25 ± 0.125	1750 ± 15
orange	2.00 ± 0.125	1720 ± 15
brown	1.75 ± 0.125	1690 ± 15
yellow	1.50 ± 0.125	1660 ± 15
green	1.25 ± 0.125	1630 ± 15
olive green	1.00 ± 0.125	1600 ± 15
cyan	0.75 ± 0.125	1570 ± 15
royal blue	0.50 ± 0.125	1540 ± 15
blue	0.25 ± 0.125	1510 ± 15
black	< 0.125	< 1495

Table 13.1 Color coded scale of the velocity and attenuation coefficient

reflected signal from the surface of the glass plate with no specimen present.

The velocity of ultrasound in the tissue specimen is calculated from the difference in phase between the signals reflected from the anterior and posterior surfaces of the specimen in comparison with the phase of the reflected signal from the surface of the glass plate in the coupling medium alone. The thickness of specimen is determined by an interference method involving the wave reflected from the anterior surface and that from the posterior surface of the tissue specimen.[4)]

The target tissue elements are determined quantitatively with the specific tissue elements identified on the two-dimensional optical image.

The attenuation and velocity data are exhibited via a color scale which has 11 steps of intervals of 0.25 dB/mm/MHz for the attenuation rate and of intervals of 30m/s for the sound speed, Table 13.1. Red shows the area where the rate of attenuation is greatest and the velocity fastest, while black shows the area where the rate of attenuation is least and the velocity slowest.[5)]

13.3 Attenuation and Velocity of Ultrasound in Fibrotic Tissue in Myocardium for Cases of DCM, HCM, and CS

The materials in this study are 15 cases with dilated cardiomyopathy(DCM), hypertrophic cardiomyopathy (HCM), and cardiac sarcoidosis (CS). Figure 13.3 shows the optical and acoustic microscopic images of a case of DCM. The optical image (left) is stained by Elastic Masson's trichrome method, in which green represents collagenous fibers and deep purple elastic fibers. The middle and right side figures are the acoustic microscopic images of attenuation and velocity,

13. *Acoustic Properties of the Fibrous Tissue* 235

(ATTENUATION 145)   (SPEED 130)

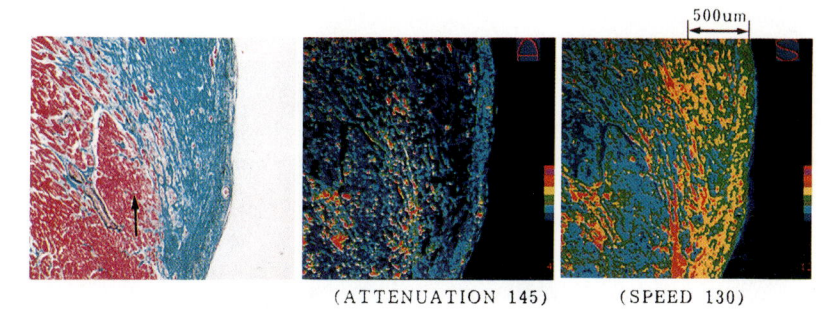

Fig. 13.3 Optical and Acoustical microscopic images of the abnormal area of the left ventricular wall in a case of dilated cardiomyopathy. The black arrow identifies the degenerated area. left: acoustic image; middle and right: optical image

(ATTENUATION 150MHz)   SPEED 130MHz

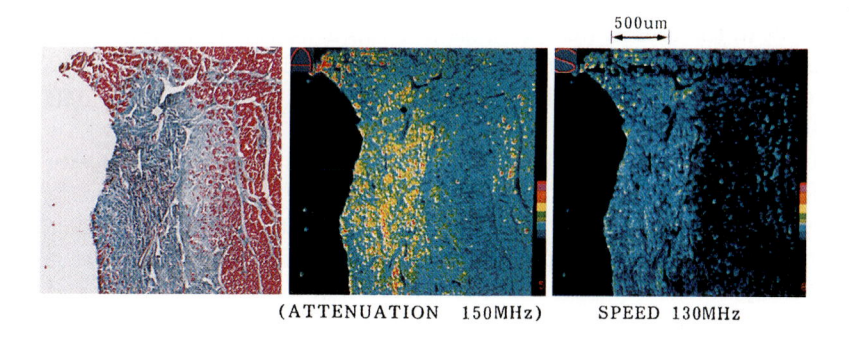

Fig. 13.4 Optical and acoustical microscopic images of the abnormal area of the left ventricular wall for a case of hypertrophic cardiomyopathy

respectively.

The optical image shows a marked increment of fibrous tissue in the subendocardium and perivascular area and in the myocardial area adjacent to the thickened fibrous endocardium. The degenerated area, as shown by the black arrow in the figure, is observed near the fibrotic changes. In the acoustic images, the corresponding areas of fibrotic myocardium are yellow to magenta and, in some cases, fibrotic tissue ranges from yellow to red. These results indicate that the velocity in the fibrotic tissue in DCM is 1600 m/s to 1750 m/s. As regards the attenuation image, the fibrotic tissue shows cyan to blue, implying that the attenuation coefficient is in the range of 0.75 to 1.00 dB/mm/MHz and is occasionally 1.5 to 1.75 dB/mm/MHz.

Figure 13.4 is a case of HCM. In the attenuation figure, yellow to orange is observed in the middle of the scar-like fibrous tissue. The dominant colors on the attenuation and velocity images are cyan to green. Therefore, the attenuation

ATTENUATION 130MHz   SPEED 120MHz

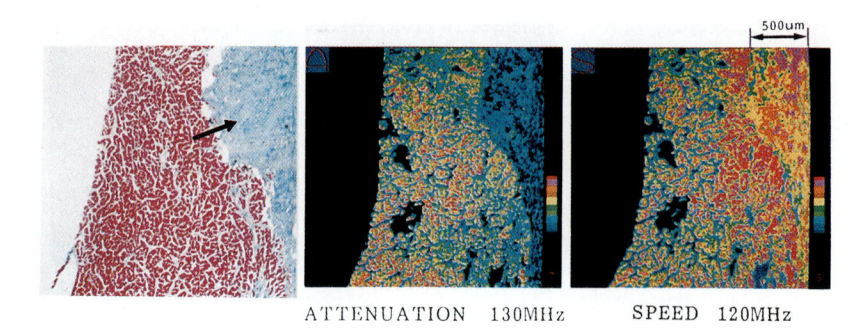

Fig. 13.5 Optical and acoustical microscopic images of the abnormal area of the left ventricular wall in a case of cardiac sarcoidosis. The glanuloma is stained light blue (←)

coefficient in the area of fibrous tissue in HCM is mostly in the range of 0.75 to 1.00 dB/mm/MHz and is occasionally 1.25 to 1.75 dB/mm/MHz. The velocity in this area is in the range of 1540 m/s to 1570 m/s and occasionally 1600 m/s to 1660 m/s.

Thus, the velocity of the fibrous tissue in HCM is about 100m/s slower than that for the case of DCM.

Figure 13.5 is a case of CS. The area of granulomatous changes in myocardium is stained light blue in the optical image. The corresponding area in acoustical image is represented by yellow and brown areas in the velocity image, and royal blue to cyan in the attenuation image. Therefore, in the area of granulomatous changes, the velocity is in the range from 1660 to 1690 m/s, and occasionally 1750 m/s, and the attenuation coefficient is as small as 0.25 to 0.75 dB/mm/MHz.

Table 13.2 is a summary of the measurements of velocity and attenuation coefficient of surviving myocardium and abnormal fibrous tissue in myocardium.

13.4 Shape and Structure of the Boundary Surface between Normal and Abnormal Tissues in Myocardium

As observed in Fig. 13.1 for the case of HCM, thick fibrous tissue is frequently observed in the endocardial area, and focal degeneration and fibrosis of the myocardium, stained blue, appear in the myocardial area. In such cases, wide strong echoes and large speckle echoes are also observed at the corresponding areas in the endocardium and myocardium on respective echocardiograms. In cases with DCM, as demonstrated in Fig. 13.6, thickening and fibrotic changes cover the broad area of the inner surface of the ventricle. The fibrotic tissue increases within the myocardium in a linear arrangement and in a reticular

13. Acoustic Properties of the Fibrous Tissue

	Tissue Element	Attenuation (dB/mm/MHz) Range	Mean	Velocity (m/s) (c) Range	Mean	ρ	ρ C (=Z)
Normal	Myocard	0.7~1.4	1.07	1570~1660	1618	1.082	1.751
	Blood	1.6~2.25 (blood cell)	1.70	1550~1620	1583	1.055	1.672
DCM	Degen.	0.5~0.75 (~1.0)	0.63 (0.75)	1510~1570	1560	1.074	1.675
	L (fib)	0.25~0.75	0.5	1570~1630	1630	1.118	1.822
	D (fib)	0.75~1.00 (~1.5)	1.13 (1.17)	1600~1660 (~1750)	1712	1.118	1.914
	Elast. f.	0.75~1.00 (~1.75)	1.23 (1.25)	1570~1630	1590	1.118	1.778
HCM	Degen.	0.5~0.75	0.63	1510~1540 (~1570)	1555	1.074	1.67
	L (fib)	0.75~1.00	0.86	1540~1570 (~1630)	1602	1.118	1.791
	D (fib)	0.75~1.25 (~2.0)	1.23 (1.27)	1540~1630 (~1660)	1630	1.118	1.822
	Interstit. T.	0.5~0.75	0.63	1570~1630	1600		
CS	Degen.	0.5~0.75	0.7	1540~1570	1566	1.074	1.682
	L (fib)	0.25~0.5	0.45	1630~1660	1650	1.118	1.845
	D (fib)	0.75~1.0	0.78	1690~1750	1702	1.118	1.902
	Cell Inf.	0.75~1.5	1.0	1660~1765	1715		

Table 13.2 Results of measurement of velocity and attenuation of the normal and abnormal tissue elements. The velocity of blood was measured by the multireflection method in the fresh blood. Degen: degeneration; L: fibrous tissue which stains light green; D: fibrous tissue which stains dark green; Cell Inf: cell infiltration; ρ: density

(ATTENUATION 145)   SPEED 130MHz

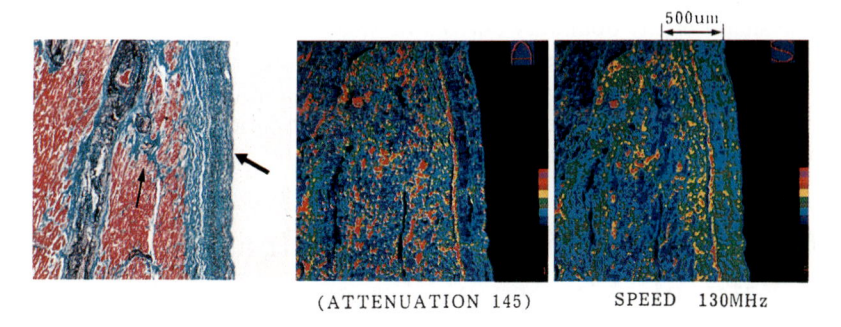

Fig. 13.6 Optical and acoustical microscopic images at the area of fibrotic changes in the left ventricular wall in a case of DCM. Fibrotic thickening of the endocardium (←), perivalscular fibrosis, and linear arrangement of the fibrotic tissue invaded into the myocardium (←) are observed

Fig. 13.7 Schematic illustration of the shape and structure of the fibrous tissue appearing in myocardium in DCM, HCM. F: Fibrous tissue; A: artery

pattern in the optical microscopic image (refer to the right lower picture of Fig. 13.1).

As understood from the optical microscopic images, the length of the fibrotic tissue appearing in the myocardium, in cases of DCM, HCM, and CS, is more than several millimeters. The size of the localized spotted fibrotic change, which is frequently observed as blue or green in the myocardium in Fig. 13.1, ranged from several millimeters to more than 10 mm and was more than several times as broad as that of the myocardial fiber.

On the other hand, when 3MHz pulsed ultrasound with a 0.5 mm wave train length is used, as in B-mode focused ultrasound echocardiography, the effective ultrasonic beam width in the tissue will be less than 3 mm at the focal area. Consequently, when ultrasound is used in the pulse reflection mode and the target tissue to be measured is placed at the focal area, it must be considered that the shape of the boundary surface between normal and abnormal tissues, or between two abnormal tissues, is more similar to that of polyhedral reflectors than to that of spherical reflectors, as illustrated in Fig.7.[6)7)]

Further, it can be assumed that the tissue boundaries, which occur in the pathological state, function as plane reflectors.

13.5 Relationships Between Changes in Quality of Abnormal Tissue and Intensity of Reflection or Reflectivity at the Abnormal Tissue Boundaries

Providing that the assumption, mentioned in the previous section is valid, the sound pressure reflection coefficient r and that of the intensity reflection param-

13. Acoustic Properties of the Fibrous Tissue

eter Rl can be represented by:

$$r = \frac{P_r}{P_o} = \frac{Z_1 - Z_2}{Z_1 + Z_2}$$
$$RI = 10\log r^2 = 20\log \frac{Z_1 - Z_2}{Z_1 + Z_2} \tag{13.1}$$

where Z_1 and Z_2 are the acoustic impedances of contiguous tissue elements, and P_r and P_o are the sound pressure amplitudes of the reflected and incident waves, respectively.

In echocardiography, the attenuation of ultrasound during transmission from the chest surface to the heart is significant and a method for correction of this loss of ultrasound energy must be incorporated into the measurement of echo intensity. It is, therefore, convenient to define a standard reflector for the relative measurement of r and RI in living human heart. For this study, the boundary between the heart wall and blood, assuming a flat (planar) endocardial surface, was defined as the standard reference reflector because although the echo from normal endocardium is very weak, it can be recorded for nearly all patients.

The intensity of the echo presumed to be reflected from various tissue boundaries was calculated and obtained in comparison with the predicted intensity of the standard reference echo from the endocardial surface. These calculations are shown in the RI column of Table 13.3.

As understood from Table 13.3, when reflections at a boundary between two pathological tissues are taken into consideration, the presumed reflection level ranges from about -23.5 dB to -64.4 dB. The average value of the intensity reflection parameter is -32.3 dB for DCM, -34.3 dB for HCM, and -31.1 dB for CS. There was little disease dependent difference in the intensity reflection parameter. Except for reflections at the boundary between degenerated tissue(Deg) and blood(B), the average value of this parameter exceeded -40 dB at most boundaries, viz., -30.1 dB in DCM, -30.5 dB in HMC and -28.9 dB in CS. These results indicated that the intensity reflection parameter at the tissue boundary is greater than that of the boundary between normal myocardium and blood.

Table 13.3 also shows the values of reflectivity of an interface between the normal myocardium and abnormal tissue with pathological changes in comparison with the reference value, viz., the reflectivity at the interface between the normal myocardium and blood (Ref, RI). The reflectivity thus calculated is in the range from -63.2 dB to 18.6 dB because the ultrasound travels in both directions, going and returning, in echocardiography.

The reflectivity is seen to be low at the interface between normal myocardium and degenerated myocardium with vacuolated change, and with fibrous tissue L, which stains light green in the optical image, viz., in the range of -2.4 to 2.4 dB. On the contrary, high level reflectivity, of the order of 10 to 11 dB, is found at the interface between the normal myocardium and fibrous tissue D, which stains dark blue in the optical image.

Thus, it can be stated that the collagenous fibrous tissue in myocardium is

		Z_1 — Z_2	γ	$RI=20\log_{10}\gamma$	dB	RI average (dB)
N	M-B	1.751-1.672	0.023	-32.8	0	
C D	D-Deg	1.914-1.675	0.0666	-23.5	9.3	
	D-B	1.914-1.672	0.0609	-24.4	8.4	
	D-M	1.914-1.751	0.0444	-27.0	5.8	
	L-B	1.822-1.672	0.0429	-27.3	5.5	
	L-Deg	1.822-1.675	0.0420	-27.5	5.3	
M	D-E	1.914-1.778	0.0368	-28.7	4.1	
	E-B	1.778-1.672	0.0307	-30.2	2.6	
	E-Deg	1.778-1.675	0.0298	-30.5	2.3	
	L-E	1.822-1.778	0.0284	-30.9	1.9	
	D-L	1.914-1.822	0.0246	-32.2	0.6	
	Deg-M	1.675-1.751	0.0222	-33.1	-0.3	
	L-M	1.822-1.751	0.0200	-34.0	-1.2	
	E-M	1.778-1.751	0.0077	-42.2	-9.4	
	Deg-B	1.675-1.672	0.0009	-61.0	-28.2	
C H	D-Deg	1.822-1.670	0.0614	-24.2	8.6	
	D-B	1.822-1.672	0.0609	-24.4	8.4	
M	L-Deg	1.791-1.670	0.0462	-26.7	6.1	
	L-B	1.791-1.672	0.0429	-27.4	5.4	
	Deg-M	1.670-1.751	0.0240	-32.4	0.4	
	D-M	1.822-1.751	0.0200	-34.0	-1.2	
	D-L	1.822-1.791	0.0152	-36.4	-3.6	
	L-M	1.791-1.751	0.0110	-39.2	-6.4	
	Deg-B	1.672-1.670	0.0006	-64.4	-31.6	
S a r c o i d	D-B	1.902-1.672	0.0644	-23.8	9.0	
	D-Deg	1.902-1.682	0.0614	-24.2	8.6	
	L-B	1.845-1.672	0.0492	-26.2	6.6	
	L-Deg	1.845-1.682	0.0422	-26.7	6.1	
	D-M	1.902-1.751	0.0410	-27.8	5.0	
	L-M	1.845-1.751	0.0260	-31.6	1.2	
	Deg-M	1.682-1.751	0.0200	-34.6	-1.2	
	D-L	1.902-1.845	0.0152	-36.4	-3.6	
	Deg-B	1.682-1.672	0.0030	-50.6	-17.8	

Table 13.3 Calculated data of acoustic impedance, reflection coefficient (r) and intensity reflection parameter (RI). Z1, Z2 : Acoustic impedances; M: normal myocardium B : blood; Deg : degenerated; D : fibrous tissue which stains dark green;   L : fibrous tissue which stains light green;   E : Elastic tissue

13. Acoustic Properties of the Fibrous Tissue

responsible for much stronger reflections than occurs at the boundaries within the normal myocardium.

13.6 Detectability of Echoes at the Tissue Boundary by the Use of Commercially Available Echocardiographic Instruments

From the practical view point, it is necessary to investigate whether the echo at a tissue boundary with minimum level of reflectivity can be detected by a commercially available echocardiography instruments. Thus, as part of this study, a new experimental model of a reflection plane which produced a very weak reflection, with a reflectivity of -30 dB below that of the reference reflection plane, was developed. This method employed the boundary between two water layers at slightly different temperatures.

The specific acoustic impedance of water at different temperatures was determined from data cited in the literature.[8] Then, the reflection coefficient and reflectivity at the boundary between two water layers, which are at different temperatures, could be predicted from the acoustic impedances so obtained.

Figure 13.8 shows graphically the correlation between reflectivity predicted (ordinate) and water temperature (abscissa).

This figure exhibits the changes of the predicted reflectivity of the ideal reflection plane, when water at various temperature levels is poured rapidly into the water bath at a fixed temperature. The predicted reflectivity was calculated under the assumption that water at varying temperatures is poured into a water bath at 0, 5, 15, 35, 45, 55 and 60°C, respectively.

An ideal reflection plane, having varying levels of reflectivity, can be constructed by changing the temperature of water to be poured into the water bath. In order to test this method experimentally, a reflection plane with the small reflectivity of -50 dB to -60 dB, respectively, was demonstrated by pouring water at 40°C to 70°C, respectively, into a water bath filled with water at 35°C, from a quadrangular plastic infusion tube installed on the inner wall of the water bath. Water was introduced by a piston pump as quickly as could be accomplished manually.

The echo level thus produced was about -50 dB to -55 dB, as illustrated in Fig. 13.8 by the symbol (●).

The echo produced at the boundary between two water layers at different temperature was displayed by two-dimensional echocardiography as two parallel linear echoes, as shown in Fig. 13.9. To produce weak echoes of about -60 dB to about -70 dB, water at 42°C to 70°C was poured quickly into a water bath filled with water at 45°C.

Temperatures plotted and shown by the double circle on Fig. 13.8 are associated with good quality echoes detected on the CRT screen.

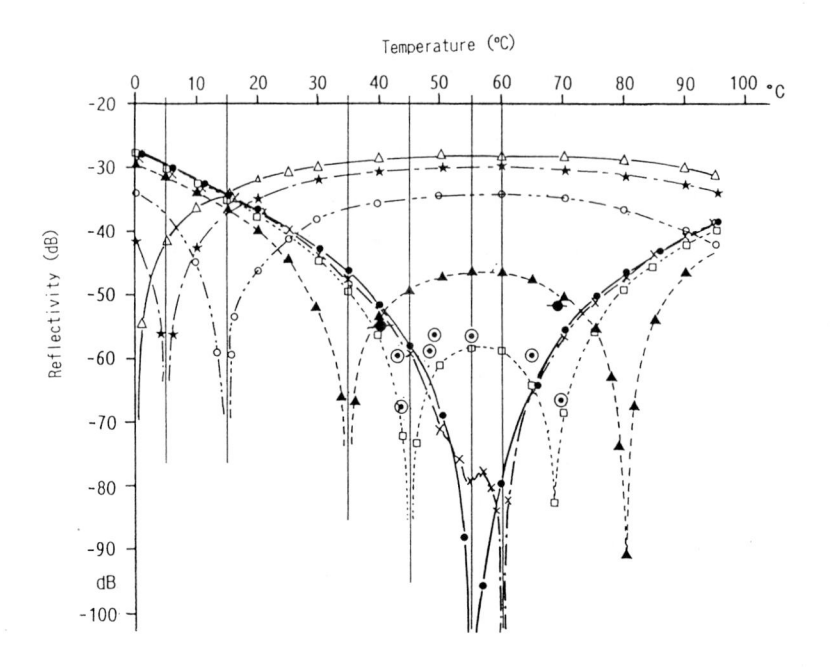

Fig. 13.8 Graphic display of the correlation between reflectivity (ordinate) and infused water temperature (abscissa). ● and ◎ show actual measurement data. △: Water temperature in the water bath is 0 ℃. ★: 5 ℃ , ○: 15 ℃ , ▲: 35 ℃ □: 45 ℃ , ●: 55 ℃ , ×: 60 ℃

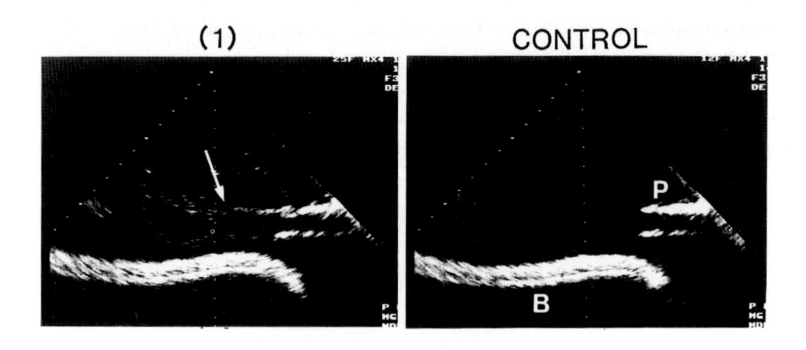

Fig. 13.9 Two-dimensional echogarm of the boundary echo with the smallest reflectivity. Two parallel echoes (white arrow) were obtained at the boundary between 45 ℃ water in the water bath and the infused water of various temperature. P: the echoes of the infusion tube; B: bottom echo

13.7 Conclusion

It is thus concluded that weak echoes in the range of -50 dB to -64 dB, that are frequently produced at the boundary between degenerated tissue and blood, can be detected by commercially available echocardiographic instruments. Further, abnormal echoes occurring in the myocardium can be identified and evaluated quantitatively, during clinical echocardiographic examination, by taking into account the attenuation loss experienced during transmission of the reflected wave from the abnormal tissue boundary to the chest surface.

References

1. M. Tanaka, et al. : Non-invasive estimation by cross sectional echocardiography of myocardial damage in cardiomopathy. Br. Heart J. 53:137~152, 1985.

2. M. Tanaka, et al. : Development of the acoustic microscope for the medical and biological use and its medical application, Kokenshi,37: 377~387, 1985. (in Japanese)

3. M. Tanaka, : Usefulness of ultrasonic image in the medical field, Acoustical Imaging, 17: 453~466, 1989.

4. H. Okawai, M. Tanaka and F. Dunn: Non-contact acoustic method for the simultaneous measurement of thickness and acoustic properties of biological tissues. Ultrasonics 28: 401~410, 1990.

5. H. Okawai, M. Tanaka and N. Chubachi: Two-dimensional quantitative display in color scales for acoustic properties of a tissue by using the scanning acoustic microscope. Ultrasonic Technology, 1987; ed by Kohji Toda, 15~24, 1987. MYU Research, Tokyo.

6. T. Takahashi and J. Matsumoto: Topological analysis of the chronic liver diseases: Liver; 18: 408~418, 1977. (in Japanese)

7. T. Takahashi: The development of liver cirrhosis from the viewpoint of lobular architecture. Endoscopia Digestive. 2(7): 839~846, 1990. (in Japanese)

8. Hand-Book of Ultrasonic Technology, p 1195, ed. by J. Saneyoshi, Y. Kikuchi, O. Nomoto, Nikkan-Kogyo press, Tokyo, 1960. Nov.

Closing Remarks

NITTA: Thank you, Dr. Thijssen and Dr. Chubachi. At the end of the Sendai Symposium on Ultrasonic Tissue Characterization 1994, Professor Chubachi will make closing remarks.

CHUBACHI: It is our great pleasure to have all the experts in the field of tissue characterization from around the world here in Sendai, one of the places where ultrasonic diagnostics originated. I think it is a most wonderful opportunity to have this kind of meeting in Sendai, and I should like to consider that the late Professor Y. Kikuchi would have been looking forward to this opportunity, if he were alive today.

Since early this morning we've been discussing, in a very enthusiastic manner, some very most fascinating topics. I am very happy to say that you look satisfied to have participated in this symposium, although you may be exhausted.

Last night at the welcome party, Dr. Hill said in his speech that he used to encourage young researchers in England to have an interest in tissue characterization. And today, I think many people mentioned that this science is becoming more and more interesting. This field of research is very difficult to cultivate, but it is really going to be a science, or at the very least, it's developing rapidly into a new science. So, we are very glad to have many young Japanese researchers attending this symposium and I hope all the young researchers have been inspired with many good ideas and food for thought.

Most of the speakers today are around or over the age of 60, I may say. We've been working hard over the past 20 years, and it has become evident that this science is becoming more and more important, so we need the help of the young researchers to develop the field. I think it is also most fortunate for you young researchers in Japan to be able to participate in this symposium with us today.

In the first talk given by Dr. Hill, he indicated, by showing a picture of the Tower of Babel, it is very difficult for researchers to communicate with each other, owing to the various different problems. Today, as far as language problems are concerned, we have had excellent interpreters, I think by whose help we could overcome language barriers between Japanese and English in our very constructive discussions. In particular, I am very glad to see that Professor Tanaka seemed to be truly enjoying the discussion all the way through.

---

*N..Chubachi, Ph.D.

I believe that tissue characterization is going to contribute to the welfare of people in the 21st century. So for the augmentation of human happiness, I wish that all the scientists in the world will become interested in this field and work together, so that God, who destroyed the Tower of Babel, will in turn pleasingly help to build our Ultrasono-Tower.

I should like to express my sincere thanks for having such a large attendance in spite of today being Sunday, and for making it such a fruitful meeting. All members of the organizing committee are very happy and very thankful to all of you.

The proceeding of the meeting will be published, of course to be reviewed by you first, so that the fruit of this symposium will be shared not only by us here, but also by all the other young researchers who could not attend this meeting.

We hope there will be more and more progress in the research and the scientific developments of this field. And I hope that the Sendai symposium will serve as a milestone in that regard.

Thank you very much again. (The closing remarks were given in Japanese to encourage many Japanese young participants.)

Index

absorption, 26, 181
absorption coefficient, 25, 26, 181
acceleration pickup, 130
acoustical microscopy, 201
acoustic characteristics, 232
acoustic element, 176
acoustic impedance, 239, 240
acoustic microscope, 185
acoustic microscope system, 173
acoustic property, 171, 180
acoustic spectrographic imaging, 77
adaptive filtering, 84
adiabatic volume compressibility, 181
adiabatic volume elasticity, 181
A-mode profile, 194
amplitude detector, 176
amplitude image, 217
aorta, 126
aortic valve, 135
aortic wall, 125
arterial wall, 125
arteriosclerosis, 126, 132
artificial neural network, 86
ascending aorta, 132
attenuation, 9, 181, 202, 212
attenuation coefficient, 4, 77, 78, 234, 235
attenuation constant, 190, 218
attenuation rate, 232
average differential scattering cross section, 34
axis, 58

backscattering, 77
backscattering coefficient, 77
backscattering cross section, 4

backscattering spectra, 78
backscatter intensity, 78
basal cell carcinoma, 206
biological tissue specimen, 180
breast cancer, 112, 114
breast carcinoma, 114
breast tissue, 109
B-scan texture, 4
bulk backscattering cross-section, 9
bulk modulus, 69
bulk property, 4

calf liver, 37
carbon tetrachloride, 65
cardiac sarcoidosis, 234
choroidal tumor, 84
cirrhotic liver, 66, 69
classical absorption, 181
C-mode image, 173
collagen, 145
collagenous fiber, 234
collagenous fibrous tissue, 239
color coded scale, 234
color shading, 193
compressibility, 39
concave circular transmitter, 60
concave sapphire lens, 232
contact method, 180
contiguous tissue element, 239
copper sulfate solution, 66
correction factor, 183
coupling liquid, 180, 201
coupling medium, 176, 217
Cramér-Rao lower bound, 103
cross-correlation, 101
C-scan, 232

cyclic variation of backscatter, 144

degenerated area, 235
degenerated tissue, 239
density, 39
density measurement, 66
dermatology, 201
dermis, 210
differentiated adenocarcinoma, 219
dilated cardiomyopathy, 234
disk transmitter, 57, 60
dispersion, 190
Doppler, 96
Doppler effect, 126

echo amplitude, 210
echocardiography, 238, 239
echographic imaging, 217
echo level, 241
ejection phase, 126
elastic fiber, 234
elasticity, 95, 125
Elastic Masson's trichrome method, 234
elastic modulus, 97
elastic tissue, 240
elastic wave velocity, 97
elastogram, 103, 109
elastography, 99
endocardial area, 236
endoscopic ultrasonography, 218
epidermis, 210
external excitation, 97

fat content, 69
fatty degeneration, 69
fatty liver, 19, 23, 66
fibrosis, 69
fibrotic myocardium, 235
fibrotic thickening, 237
fibrotic tissue, 235
flexor muscle, 114
focal degeneration, 236
formalin fixed, 185
fractal analysis, 84

fractal dimension, 81
frequency characteristic pattern, 188
frequency characteristics, 181
frequency dependence, 37, 77
frequency-dependent attenuation coefficient, 78
frequency exponent, 188
frequency spectra, 4
frequency-time analysis, 65
fringe shift, 179

gastric cancer, 217, 218
granulomatous change, 236

heart muscle, 125
heart tissue, 66
heart wall, 125
hematoxylin-eosin stain, 206
homogeneous media, 53
human eye, 84
hypertrophic cardiomyopathy, 234

ideal reflection plane, 241
immiscible liquid model, 69
incident wave vector, 31
inhomogeneous material, 180
inhomogeneous media, 53
insulin-dependent diabetics, 146
integrated backscatter, 139–141
intensity reflection parameter, 239, 240
interferogram, 179
interrogation rates, 8
interventricular septum, 126
irradiation power, 26
ischemia, 141
ischemic heart disease, 145

L2 mean, 81, 83, 84
lateral resolution, 173
light microscope, 185
Lord Rayleigh, 53

macroscopic bulk property, 171
malignant melanoma, 206
mammography, 109

Index

mechanically scanned acoustic microscope, 172
mechanical scanner, 173
microscope resolution, 232
microscopic property, 171
microtome, 185
Minkovsky dimension, 85
M-mode, 126
mucosa, 219
multiparameter, 75
multiple parameter confusion, 8
myocardial anisotropy, 150
myocardial fiber, 238
myocardial fibrosis, 126
myocardial tissue characterization, 139

non-contact measurement, 63
non-contact method, 180
non-dispersive, 188
nonlinear dependence of absorption, 23
nonlinearity parameter, 4, 9

optical microscopy, 202
optimum frequency, 190

palpable elastic property, 95
palpation, 95
papillary adenocarcinoma, 219
paraffin embedded, 185
parametric imaging, 76
perivalscular fibrosis, 237
phase image, 217
phase shift, 179, 190
pig fat, 20
pig liver, 20, 43
plane transmitter, 54
point-spread function, 80
Poisson's ratio, 97, 98
polyhedral reflector, 238
power spectra, 126
precision, 6, 77
predicted reflectivity, 241
psoriasis, 202
pulse wave velocity, 125, 126

qualitative texture analysis, 139
quantitative texture analysis, 139
quantitative two-dimensional display, 190
quantitative ultrasonology, 3

radio-frequency signal, 76
rat liver, 63
Rayleigh's expression, 62
Rayleigh probability distribution function, 84
receiver operating characteristic, 146
reconstructive interpolation method, 102, 103
reference reflection plane, 241
reflection coefficient, 238, 240
reflectivity, 239
relative contrast, 6
renal cell carcinoma, 217, 228
renal tumor, 228
resolution, 77
ring function, 55, 57, 58

sapphire lens, 201
scanning acoustical microscope, 201
scanning acoustic microscope, 171, 217, 233
scanning acoustic microscopy, 231
scanning laser acoustic microscope, 172
scar-like fibrous tissue, 235
scattering, 31, 181
scattering strength, 82
scattering vector, 31, 33
second-order statistics, 84
shear elastic modulus, 4
shear viscosity, 181
signal-to-noise ratio, 81
signet-ring cell carcinoma, 219
soleus muscle, 114
sound field, 54
sound speed, 4, 63, 190, 218, 232
spatial resolution, 7
specific acoustic impedance, 241
specimen thickness, 180, 190

speckle, 81
speckle echo, 236
speckle-like echo, 231
spectral power decomposition, 40
spectral shift attenuation estimation
   method, 77
spherical reflector, 238
spherical transmitter, 56
spotted fibrotic change, 238
standard reflector, 239
strain, 97
strain distribution, 107
strain tensor, 98
stress, 97
sub-Rayleigh statistics, 82

target tissue element, 234
temperature rise, 19, 23
texture parameter imaging, 81
thermocouple, 23
thermocouple probe, 172
thickness of specimen, 234
thoracic aorta, 132
time domain correlation, 97
tissue boundary, 239, 241
tissue characterization, 75
tissue element, 232
tissue morphology, 43
tissue thickness, 181
topological dimension, 85
transfer function, 132
two-dimensional echocardiography, 241

ultrasonic absorption microscope, 172
ultrasonic beam width, 232, 238
ultrasonic tissue characterization, 3,
   217
ultrasonic transducer, 174
ultrasound absorption, 19

vacuolated change, 239
velocity, 234
viscous action, 26
V-Z curve method, 173

wavespace, 43

Young's modulus, 97, 99, 130